"*Global Burning* is as powerful as it is succinct. Eve Darian-Smith writes with urgent clarity and conceptual richness as she grapples with some of the most pressing issues of our times. *Global Burning* is a very teachable book—truly interdisciplinary and international in reach."
—ROB NIXON, author of *Slow Violence and the Environmentalism of the Poor*

"*Global Burning* is a brilliant analysis of how a range of anti-democratic trends can be viewed through the lens of catastrophic wildfires across the globe. If you want to understand how to analyze and become involved in a politics of collective resistance aimed at saving both the planet and democracy itself, this is the book to read."
—HENRY GIROUX, author of *Race, Politics, and Pandemic Pedagogy*

"The threat of extinction is real and immediate, but Eve Darian-Smith rightfully warns that it cannot be effectively thwarted unless we link the fight for environmental survival with the struggles against global, class, racial, and gender inequalities. A persuasive, solidly documented work."
—WALDEN BELLO, co-founder of Focus on the Global South and recipient of the Right Livelihood Award

GLOBAL BURNING

Global Burning

Rising Antidemocracy and the Climate Crisis

EVE DARIAN-SMITH

STANFORD UNIVERSITY PRESS
Stanford, California

STANFORD UNIVERSITY PRESS
Stanford, California

Printed in the United States of America on acid-free, archival-quality paper

ISBN 9781503631083 (paper)
ISBN 9781503631465 (ebook)
Library of Congress Control Number 2021055739

CIP data available upon request

Cover design: Rob Ehle
Cover photo: Mural in Chivary, Peru. Eric Hirsch
Text design: Kevin Barret Kane
Typeset by Motto Publishing Services in 10/14 ITC Galliard Pro

*To my dearest mum, Pammie, who slowly faded
away across the ocean amid massive bushfires
and then months of lockdown, as I wrote this book.*

CONTENTS

FIGURES

PREFACE

I began working on this book in January 2020 amid great global concerns about the Australian bushfires that had been burning for months across much of the country. I live in Southern California, but I am an Australian and was very anxious about the news. At the top of my street, near the local school, I saw some kids selling lemonade to support kangaroos and koalas recovering from burns caused by the massive bushfires. I had just heard that three visiting firefighters from the United States had died in a plane crash while battling the flames. The lemonade stand and fate of the unlucky firemen brought the Australian disaster closer to home. What was happening on the other side of the world was making people jumpy. In California preparing for fire evacuations has become standard procedure, even among Hollywood's rich and famous. Like most people, I know to move fast when given the signal, and bottles of water and snacks are squirreled away in my car. But news of the Australian fires was capturing people's imaginations in a new way. Ever-watchful meteorologists were telling us that smoke from the Australian fires could be detected over much of South America, including parts of Argentina, Uruguay, and Rio Grande do Sul in Brazil. Bizarrely, this smoke was merging with thick haze produced by tens of thousands of fires still burning across the Amazon jungle that had displaced many villagers and Indigenous communities, and devastated huge swaths of rainforest throughout 2019. Untold numbers of people, animals, birds, and plants had died in the flames, and unique ecosystems had been destroyed forever. From outer space, satellites confirmed what millions of people were experiencing on the ground in places as distant as Australia, California, and Brazil: our planet is on fire.

However, as the month of January progressed, the news cycle changed, shifting from stories about out-of-control wildfires to an emerging health crisis in China with the outbreak of the COVID-19 virus. Within weeks, the world's focus had entirely turned to the spreading disease. The World Health Organization declared a global pandemic on March 11, 2020, as the number of deaths steadily rose in Italy and Spain. The initially skeptical leaders of Britain and the United States were forced to acknowledge that their countries were facing a serious public health crisis. Many governments declared a national state of emergency. Around the world many people returned to their place of residence, and entire labor, business, and educational sectors began preparing to work remotely. As tens of millions of people learned how to shelter in place, the global economy slowly ground to a halt. Fears of economic recession dominated corporate-owned news media. More pressingly, people around the world contemplated how to secure food and shelter in times of mass layoffs, rising unemployment, and underfunded health systems. It seemed the whole world was going mad as visions of apocalyptic doom saturated social media and people engaged with the proximity of death.

As I put the finishing touches to this book in August 2021, the pandemic still rules, with Brazil and particularly India devastated by surging cases and lack of adequate oxygen and hospital beds. It is impossible to accurately predict how many hundreds of thousands of people will die of COVID-19. Even with the rolling out of numerous vaccines in the first half of 2021, how quickly the global population will receive medicine to stop the mutating virus remains a big question. Vaccine nationalism is a new concept that is dominating the global political economy and recalibrating asymmetrical power relations between rich and poor nations. The United States, for instance, has been hoarding vaccines and offers third booster shots while most of the countries in Africa have not vaccinated even 10 percent of their populations. No one knows how the pandemic's impact will play out over the coming months and possibly years, but everyone knows it will harm marginalized communities of color and poorer countries disproportionately.

What is certain is that climate change—which both contributes to the pandemic and exacerbates its unequal impacts—looms more starkly than ever before as a coming crisis. In August 2021, the United Nations issued a very significant scientific report—what has been called "a code red for humanity"—arguing for an immediate global reduction in greenhouse gas emissions to stall inevitable planetary warming.[1] However, firefighters remain periodically in lockdown under quarantine, temperatures continue to rise, and out-of-control fires continue to burn. More insidiously, in countries such as the United States, Brazil, and Australia, environmental protections have been rolled back as societies focus on getting their economy back on track. Even with the ousting of Donald Trump in late 2020, and emerging hope with the US rejoining the Paris Agreement and making climate change a central political issue, grave concern about the planet's future prevails. Reinstating environmental laws will not be easy, with Trump loyalists continuing to deny climate science and Congress opposing reforms through legislative gridlock. Meanwhile, around the world, radical ultranationalist governments are seizing sweeping powers under the guise of dealing with a public health crisis. Clamping down on the freedom of assembly and expression, installing invasive surveillance systems, postponing elections and militarizing the police—these measures indicate antidemocratic governance is escalating. As noted by a United Nations spokes-person, "We could have a parallel epidemic of authoritarianism and repressive measures following close if not on the heels of a health epidemic."[2] How one fights climate change amid a global rise of authoritarianism becomes a central question explored in this book.

It is important to understand how we arrived at the current moment. Many of the issues dramatically revealed by the pandemic have been determining the climate crisis that has been steadily unfolding for more than forty years. These include the denial of scientific evidence, government inertia, and the mass exploitation by corporations of cheap human labor and natural resources. But the pandemic has also brought global attention through the Black Lives Matter movement to systemic racism, disproportionate impacts on people of color and the poor and marginalized, and explicit indifference by a global economic elite to human life. As noted by one commentator, "We have to con-sider COVID-19 a test case for the climate crisis that is on the horizon, and it pales in comparison to the latter."[3] Adds another, "Our world cannot be in denial of the climate crisis any longer. COVID-19 is devastating thousands of lives and threatening mil-lions, but the impacts of climate change are endangering the lives and livelihoods of billions of people."[4]

In many ways the world's response to COVID-19 and its af-termath is a harbinger of our collective futures as well as a re-minder of how we got to the environmental crisis in the first place. Amid the unfolding short- and long-term challenges of the pandemic, this book is a call not to lose sight of the much bigger catastrophe before us: climate change and its imminent threat to all species, including human beings. We must constantly remind ourselves that there is no vaccine for water shortages, toxic air pollution, and flames licking at the front door.

Eve Darian-Smith
Southern California, 2021

GLOBAL BURNING

1

Fire as Omen

Introduction

THIS BOOK EXPLORES the causes and consequences of catastrophic wildfires that have been burning in both the northern and southern hemispheres between 2018 and 2021. Why are these bushfires occurring, and what can be done to prevent them? What are the direct and indirect impacts of fires, and who will suffer the most in the long run? Catastrophic fires are a lens through which to explore the bigger issue of climate change. Fires are only one front of the unfolding environmental crisis facing humanity, which includes toxic pollution, megadroughts, rising seas, and the massive destruction of biodiversity around the world.[1] The Australian fires—and those faced in the Amazon, California, and other parts of the world—can be interpreted as an omen of ecological collapse as temperatures rise, glaciers melt, and oceans warm. In this vein, some environmental activists use fire as an emblematic metaphor summing up in a single word the immediate and impossible-to-ignore dangers of climate change. In the words of Greta Thunberg, the young Swedish environmental activist and a global leader on fighting climate change, "our house is on fire."[2] Amplifying this emergency, Daniel Wildcat, a Yuchi member of the Muscogee Nation of Oklahoma and prominent environmental scholar, argues that today's era is one of "global burning."[3]

Fires can be devastating in their violence and impact. Whether you use the term *bushfires* (Australian), *wildfires* (American), or *incêndio* (Brazilian), catastrophic fires refuse to be ignored. Unlike other barometers of climate change that often go undetected by people going about their daily routines, fires demand our immediate attention. Walls of flame racing down a hillside, crackling brush and animals fleeing, winds howling, firefighters screaming for residents to get out, people racing to cars, ash and sparks swirling—these are the sights and sounds of fire. As one journalist described it, "It sounds like a freight train and as it burns towards us, spurting fireballs ahead across the tops of the trees, it is like a kind of raging beast."[4] Fires leave in their wake black landscapes of steaming earth, burned houses and structures, charred animal carcasses, buckled remains of vehicles, and thick smoky air that can last for days, weeks, or even months. It can seem like a living hell—people slowly moving dazed and zombielike back to their homes, trying to recover anything that reminds them of a time before. In the worst instances, there are searches for the remains of family, friends, and pets.

Out-of-control bushfires are just one measure of impending ecological disaster, but it is a particularly violent and frightening event that can cause death and immense suffering. Fires can quickly force mass evacuations, down powerlines and communications, and fill the air with raining ash that affects communities well beyond the fire zone itself. Wildfires operate on a different time line than most of the challenges associated with climate change. Fire's immediate danger stands in stark contrast to many other environmental disasters that can take years, if not decades, to become materially evident.[5] As Rob Nixon argues, environmental disasters such as leaking toxins, water table depletion, and damage to the ozone layer operate according to a different time frame, inflicting "slow violence," particularly on those living in the less wealthy countries of the global south.[6] Fires, in contrast,

FIGURE 1. RFS volunteers and NSW Fire and Rescue officers fight a bushfire encroaching on properties near Termeil on the Princes Highway between Bateman's Bay and Ulladulla, south of Sydney, Australia, Tuesday, December 3, 2019 (AAP Image / Dean Lewins).

present a direct threat to life, underscoring people's vulnerabilities and total dependence on others for water, shelter, and the basics of survival. Fires produce terrifying walls of flame and smoke and blackened landscapes that persist for months and years afterward. And as with other catastrophic events, the physical, emotional, and psychological scars left by bushfires can last a lifetime (Figure 1).

Fires racing with lightning speed seem to burn indiscriminately and don't care about a person's status, ethnicity, or nationality. However, although all people are vulnerable to fires, they impact some people and communities much more than others. Fires, like other "natural" disasters such as the COVID-19 pandemic, affect lower-income, marginalized, and racialized communities disproportionately. The Black Lives Matter movement brings worldwide attention to this harsh reality, underscoring intergenerational systemic racism manifesting today in extreme police brutality as well as unconcealed white supremacy. Systemic racism is very evident in the United States but also in other countries such as the UK, Brazil, and India, where the bulk of those working on the front lines in hospitals and emergency services are largely lower-income people of color, often marginalized for their religion or ethnicity. Because these frontline jobs overlay decades of underfunded health, housing, and educational services, communities of color suffer much more death and illness in comparison to wealthier white communities. These historical and structural differences are very evident, be it under pandemic or more "normal" conditions.[7]

A focus on what people of color and marginalized communities experience, rather than the discriminatory acts by whites against them, evokes a definition of *racism* proposed by geographer Ruth Wilson Gilmore. She writes that racism is "the state-sanctioned and/or legal production and exploitation of group-differentiated vulnerabilities to premature death, in distinct yet densely interconnected political geographies."[8] This definition underscores that people of color experience racism as economic and political conditions that determine their vulnerabilities to environmental degradation (such as pollutants in the air or lead in the drinking water) and ultimately their life expectancies. These vulnerabilities exist within particular geographical landscapes and can accumulate over decades, if not centuries. And they are linked to the intertwined histories of colonialism, capitalism, and slavery—what Cedric Robinson has called "racial capitalism"—that persist within our contemporary world.[9] In the context of catastrophic wildfires, it is essential to remember that fires exacerbate preexisting vulnerabilities and disproportionately impact racially, ethnically, politically, and economically marginalized people. As I discuss in this book, an overlooked dimension of systemic racial violence is systemic environmental racism.[10] This often involves people living in places more susceptible to fires, floods, spills, and disasters and less prepared in terms of equipment, transport, resources, and personnel to deal with short- and long-term negative impacts.[11]

Given that catastrophic wildfires always involve racial disparities, thinking about preventing wildfires necessarily involves thinking about ways to stop racial and social injustices as well. Most wildfires burn in remote rural areas populated by people who often can't afford high rents and the expenses of city living. When fires burn closer to cities, they typically emerge on the outskirts of urban and suburban centers, in poor neighborhoods, shantytowns, or fringe communities. When fires break out in these marginalized residential areas, firefighting resources are often not available. Adding to this resource burden, most catastrophic fires affect people whose livelihoods are directly or indirectly involved with land and related farming and agricultural practices. This means that fires also impact many small growers and Indigenous communities, who often live closely with the land. In short, people who rely on fields, forests, jungles, and rivers, and the fish, birds, plants, insects, and animals that live in those places, are disproportionately impacted by devastating fires. These people may work as farmers or crop pickers in remote rural areas, or they may work in big-city centers in restaurants or selling fruit as street vendors. Focusing on catastrophic fires is one way to better understand how climate change—of which fires are only one measure—is dramatically and disproportionately transforming a wide range of people's lives, many who live on the margins of society and often go unnoticed in mainstream consciousness and social media.

The number of out-of-control fires burning around the world is rising in frequency and intensity each year. How and why is this happening? The scientifically proven answer is that this is a result of escalating greenhouse-gas emissions and a warming planet. But this process didn't just happen on its own. In this book I argue that we must link the upswing in catastrophic wildfires to the rise of ultranationalist, antidemocratic, mostly male leaders who together exert an enormous amount of global political and economic power, even when they are not officially functioning as heads of state.[12] These leaders ignore, deny, excuse, and in some cases actively encourage the burning of animals, people, and unique ecological systems in the name of economic progress and development of industries such as mining and agriculture. These industries shore up national economies and promote national status within the international community. Significantly, it is not a coincidence that antidemocratic leaders supporting ecologically destructive industries also bring them, and their business partners, a great deal of power and money. These leaders include, among others, Vladimir Putin (Russia), Matteo Salvini (Italy), Jair Bolsonaro (Brazil), Recep Tayyip Erdoğan (Turkey), Andrej Babiš (Czech Republic), Iván Duque Márquez (Colombia), Viktor Orbán (Hungary), Janez Janša (Slovenia), Scott Morrison (Australia), Boris Johnson (UK), Rodrigo Duterte (Philippines), Jarosław Kaczyński (Poland), Benjamin Netanyahu (Israel), Narendra Modi (India), Daniel Ortega (Nicaragua), Alexander Lukashenko (Belarus), Xi Jinping (China), and Donald Trump (United States).

Raining Ash: Brazil, United States, Australia

The years 2018 through 2021 were particularly terrible years in terms of catastrophic fires. This book focuses on three fires that burned in three countries on three different continents. I could have picked other sites, given that there is no shortage of massive fires burning around the world. But I chose bushfires burning in Brazil, Australia, and California in the United States because of the scale of their devastation, and because they share certain common features in terms of why they occurred (extractive capitalism), whom they impacted most (poor and marginalized communities), and what they suggest about an emerging global pattern of devastating firestorms and the political and economic conditions that foster them (rising antidemocracy).[13]

Importantly, all three countries claim to be liberal democratic societies, with Brazil and the US being two of the biggest democracies in the world in terms of population. But as is argued in this book, common to these countries is that they are sliding toward antidemocratic governance that is increasingly aligned with the extreme right and coalesces around ideologies of ultranationalism, nativism, and white racial purity.

With respect to land mass, Australia and Brazil are similarly sized and the United States is marginally larger. However, their respective populations are very different: Australia has approximately 25 million, Brazil has 214 million, and the US has 333 million people living within their territories. Despite this variance in population size, the United States, Brazil, and Australia are ranked by the International Monetary Fund (IMF) as among the richest countries in the world (the US is ranked as number 1, Brazil as number 9, and Australia as number 14).[14] Of course, in each country there are cities and regions that are relatively very poor, such as in the northern areas of Brazil. And if one compares per-capita incomes, Brazil ranks much lower than the US and Australia.[15] Yet even with these economic discrepancies within and between the three countries, it is reasonable to expect that their governments could provide adequate firefighting equipment and resources to stop widespread death and destruction. However, the reactions to catastrophic fires in recent years prove otherwise. Millions of acres of bushland have burned, billions of animals have been incinerated, and untold numbers of human lives have been lost. Notably, between 2018 and 2020 these three countries were led by pro-business political leaders—Trump, Morrison, and Bolsonaro—who were elected to power on ultranationalist policies that lean heavily toward being antidemocratic and authoritarian. Interestingly, all three leaders are skeptical of climate-change science, calling it a hoax promoted by environmental activists and so-called biased scientists and liberal academics.

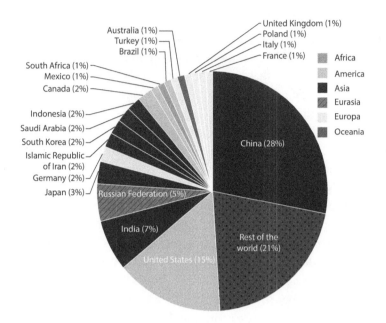

FIGURE 2. Carbon dioxide emissions by country, 2020. Courtesy of the Union of Concerned Scientists, https://www.ucsusa.org/resources /each-countrys-share-co2-emissions.

The United States, Australia, and Brazil contribute significantly to the escalation of climate change. They rank in the top fourteen nations in the world with the highest annual rates of greenhouse-gas emissions (China, the US, India, and Russia are the top four; see Figure 2). Corporations operating within them produce millions of tons of carbon dioxide pollution and greenhouse-gas emissions. What is shocking is that the governments of Brazil, Australia, and the US have politically encouraged, economically incentivized, and legally authorized companies to pollute on this massive scale. It is not surprising, then, that these three countries have led international efforts to oppose global responses to fighting climate change by reducing greenhouse-gas emissions. For instance, at the United Nations climate talks in Madrid, Spain, in

February 2020, Carlos Manuel Rodríguez, Costa Rica's minister for energy and environment, specifically blamed the United States, Brazil, and Australia for blocking efforts to tackle core issues of environmental injustice linked to global warming.[16]

Pressing questions emerge when analyzing out-of-control fires in Brazil, the United States, and Australia in recent years. What are the connections between catastrophic fires and the radically conservative governments that explicitly prioritize corporate interests and economic elites over ordinary citizens? Why did these leaders blatantly block, disregard, or dismantle environmental protection laws and the regulatory agencies that enforce them? And why is this happening when consensus on climate-change science is almost universal, and majorities of people in these countries and around the world understand the planetary crisis and believe humanity is facing some form of ecological collapse?

Changing Political Leadership in the United States

In thinking about the planet's future, does the US presidential election of Joe Biden and ousting of Donald Trump in November 2020 suggest a different trajectory? With the Democratic Party declaring climate crisis a central pillar of the new administration and rejoining the Paris Agreement after former president Trump had withdrawn the United States in 2017, many people around the world are hoping for an international reaffirmation of commitments to reduce carbon emissions and greenhouse gases. But this will require a complete about-face on environmental laws and regulations in the United States, as well as the country taking effective global leadership. Unfortunately, both aspirations face considerable challenges.

On the domestic front, a fragile balance of power governs Congress. Trump loyalists lead the Republican Party, and there is widespread opposition to environmental executive orders and legislation by Trump-appointed judges at the federal and state level. This makes it very hard to reinstate the approximately one

hundred environmental protection regulations that were dismantled during Trump's four-year term (see Appendix). Although
Biden can use his executive powers for such actions as rejoining
the Paris Agreement and pausing some leases to drill for oil and
gas on federal lands, his hands are often politically tied, preventing him from reducing the production of and widespread reliance on fossil fuels.[17] Enacting green-driven legislation requires
bipartisan support, which is difficult to achieve given the fifty-
fifty split in the Senate. Most legislation needs sixty votes for pas-
sage under the Senate's filibuster rule. Still, there is optimism
in Biden's acknowledging climate science and infusing climate
priorities across various agencies and legislative reforms. For instance, in late February 2021 the Biden administration rescinded
Trump-era guidelines that limited the consideration of long-term
carbon emissions in analyses required by the National Environmental Policy Act.

However, the real threat is that the Republican Party continues to be ruled from the sidelines by Trump, who exerts an iron
grip on the GOP. This became explicit in the dismissal of leading US Representative Liz Cheney as House Republican leader
in May 2021 after refusing to back Trump's narrative that the
2020 presidential election was "stolen." As she argued in a fiery
speech on the House floor, "Remaining silent and ignoring the
lie emboldens the liar." She went on to say, "I will not sit back
and watch in silence while others lead our party down a path that
abandons the rule of law and joins the former president's crusade
to undermine our democracy." That Cheney, the third-ranking
Republican in the House of Representatives, could be purged
from party leadership for challenging Trump is clear evidence of
the GOP's turn against the Constitution and democratic principles. Noted one commentator, "The Republican Party, by taking this action, highlights the fact that there is no room for those
who are not extremists in the party" (*New York Times*, May 7,
2021). Another commentator adds, "The Republican Party is
sliding into authoritarianism at a terrifyingly rapid clip."[18]

As Republicans gathered around the lie of electoral fraud, they furthered their attack on democracy by aggressively coordinating efforts to suppress future voter registration, targeting minority communities who typically vote for Democrats. This coordinated effort resulted in hundreds of electoral bills being proposed in mostly Republican-led states within months of Biden taking office. By June 2021 fourteen states, including Georgia and Florida, had enacted twenty-two laws extensively changing their electoral rules, and Texas was poised to follow suit. The wide-ranging "reforms" make it harder to vote by mail and to access in-person poll stations for poor and rural communities. But the new laws don't just restrict voting. They also create new procedures to subvert the voting process through racially informed redistricting that favors the GOP, as well as purging officials who have oversight in elections and who ensure accurate counting. Biden responded angrily to the news of Georgia's sweeping elections bill, calling it "an atrocity" and "Jim Crow in the 21st century." It is important to note that this latest attack on voting rights is the most recent push in a two-decade plan of voter suppression. According to former US Representative Tom Davis, a moderate Republican who represented Virginia from 1995 to 2008, "We don't just preach voter suppression, we practice it."[19]

With respect to President Biden's ability to implement environmental protections and build climate-change priorities into his infrastructure policies, he faces considerable challenges by the Trump-led Republican Party. These challenges are furthered by the public's lack of information about climate science as well as blatant lies told by Republican politicians and conservative social media outlets about green activists and renewable energy initiatives.[20] The political future is not clear, given Republicans' aggressive attack on democratic electoral processes through voter suppression, gerrymandering, partisan oversight, and fake news. There is a real possibility that Biden will be ousted in the 2024 elections and that Republicans will reverse executive orders and reinstate a fossil-fuel agenda through mining, drilling, and fracking. The return of antienvironmental national policies led by an antidemocratic Republican Party is a very real threat.

On the international front President Biden faces perhaps even bigger challenges than his domestic opposition. No matter how hard Biden asserts his global leadership through actions such as his Leaders Summit on Climate in April 2021 (where he pledged that the US would reduce global-warming pollutants by half by 2030), he must work with many world leaders who don't share his democratic principles and related environmental justice concerns. Trump's four years in office and enduring political control over the GOP emboldens far-right ideologues and authoritarian regimes. It is important to remember that Trump's promotion of violent insurrection and white supremacy—most notably in the riots at the US Capitol on January 6, 2021—were praised by autocratic leaders and radical groups around the world. This overt upswing in antidemocracy makes it harder for liberal democratic values to prevail as international norms and to feature as multilateral international goals, including a global commitment to reducing greenhouse-gas emissions to save humanity. Fueling this radical political trend are widespread propaganda and disinformation campaigns, as well as elaborate conspiracy theories such as those pushed by far-right politicians and movements such as QAnon, which circulate through domestic and global social media.[21]

Given this gloomy post-Trump, posttruth political landscape, this book's focus on the interconnections between a global rise in antidemocratic leadership and an escalating climate crisis that includes catastrophic fires remains deeply relevant and increasingly urgent, despite moments of optimism.

The Brazilian Amazon Fires

The Brazilian Amazon has been in existence for approximately 55 million years and is home to an immensely diverse plant and animal population that includes the Brazilian tapir, the giant anteater, the black dwarf porcupine, the orange-cheeked parrot, the green anaconda, the crimson-bellied parakeet, and the extraordinary Amazon pink river dolphin. Approximately 10 percent of all scientifically known birds and animals live in the rainforests, along with a further 2.5 million species of insects. The Amazon is often known as the Brazilian Amazon because approximately 60 percent of the country is covered in rainforest.

The lush rainforests of the Amazon are frequently referred to as the "lungs of the earth." But according to Carlos Nobre, a University of São Paulo climate scientist, a better way to think about "the Amazon's role is as a sink, draining heat-trapping carbon dioxide from the atmosphere."[22] Its 400 billion trees are essential for absorbing and processing tons of carbon dioxide each year produced by burning oil, coal, and other fossil fuels. It is estimated that the Amazon rainforest absorbs approximately 5 percent of the world's annual greenhouse-gas emissions. This makes the Amazon a vital resource for preventing or at least mitigating climate change. In addition, scientists now know that rainforests pull water from the ground that is then released as water vapor into the air. This process creates an atmospheric river, carrying moisture across South America and impacting rainfall patterns around the world, including California and the midwestern regions in the United States.[23] As noted by earth system scientist Paulo Brando, from the University of California, Irvine, "Climate change doesn't respect political and geographic boundaries, and what happens in the Amazon will affect the entire planet."[24]

Protecting this unique Amazonian resource against logging and deforestation had been a central pillar of Brazil's environmental policies since the early 2000s. For over two decades the country has been hailed as an international leader in conservation and efforts to combat climate change. However, in recent years Brazil's deforestation policies have been undermined by corporations seeking agricultural land. And these policies came to an abrupt halt with far-right leader Jair Bolsonaro coming to power in early

2019 on a pro-business platform advocating that rainforest protections were an obstacle to economic growth. He allowed more than eighty thousand fires to be intentionally lit by loggers and ranchers to burn enormous swaths of forest and create new commercial lands for grazing cattle, mining, and soybean production. Out-of-control fires have been mostly burning in Brazil's northwestern states of Roraima, Acre, Rondônia, and Amazonas. From August 2019 to July 2020, a total of 11,088 square kilometers (4,281 square miles) of rainforest were destroyed, a significant jump from the 7,536 square kilometers deforested in 2018.[25] As commentators have noted, there has been dramatic escalation of deforestation under Bolsonaro's leadership. "With the encouragement of the Brazilian government, deforestation of the Amazon has surged to the highest level in a decade and is now galloping along at three football fields every minute. Half of the clear-cutting and burning has occurred in protected lands." Widespread burning of the Amazon continued largely unabated throughout 2020 and into 2021, despite international condemnation by scientists, NGOs, environmental groups, and green activists.[26]

California's Record Fires in the United States

In 2018 California experienced one of its deadliest wildfires on record. The Camp Fire broke out in the remote town of Paradise in the foothills of the Sierra Nevada. The fire caused eighty-four people to die and destroyed 18,804 buildings and homes, covering an area of almost 240 square miles.[27] The Camp Fire was one of a sequence of terrible fires that burned over the previous years, all of them leaving in their wake deaths and enormous property damage (other examples include the Redwood Valley Fire in 2017, Tubbs Fire in 2017, and Mendocino Complex Fire in 2018). In 2020 another massive set of wildfires engulfed California, officially making it the worst fire season on record. Overall, 9,639 fires burned nearly 4.4 million acres, or approximately 4 percent of the state's lands. In August the crisis hit a crescendo, when 650 wildfires ignited from a rare siege of dry lightning strikes, creating massive fires. Approximately sixteen thousand firefighters from across the country and other parts of the world converged to help fight the blazes, which raged for weeks. At one point the August Complex Fire merged with the Elkhorn Fire, creating a "monster" fire that burned an area larger than the state of Rhode Island. Like fires in the Amazon and across Australia, these recent California fires destroyed unique ecosystems and plant and animal life as well as irreparably burning to the ground iconic redwoods, giant sequoias, and the unique Joshua trees. California's out-of-control wildfires also caused immense property damage that disrupted the lives of hundreds of thousands of people in densely populated areas around San Francisco, Los Angeles, and San Diego.

The reasons for California's fires are complex. As the urban historian Mike Davis wrote in an essay titled "The Case for Letting Malibu Burn," Southern California has been besieged by catastrophic wildfires throughout the nineteenth and twentieth centuries, when local Indigenous Chumash and Tongava tribes were prohibited by white settlers from doing controlled burning of the chaparral.[28] In more recent decades, land developers have pushed housing development further and further into the Santa Monica Mountains, and after each wildfire they are further incentivized to build "bigger and better" homes rather than deal with stockpiles of accumulated brush that fuel catastrophic fires in the first place. Fires, in short, are a regular occurrence for modest communities in the Santa Monica hills as well as for many wealthy white residents residing in Malibu and adjacent suburbs who own mountain holiday homes. Davis ominously noted in a postscript to his original essay, "When most of us build or buy a home, we carefully appraise the neighborhood. In Malibu the neighborhood is fire."[29]

What is unique about today's catastrophic wildfires? The answer is climate change. More specifically, it's the cumulative impacts of climate change that have been both caused and exacerbated by corporations involved in a range of loosely regulated and often environmentally risky activities such as energy production, industrial-scale agriculture, mining and drilling, and massive extraction of groundwater. Together these industries can be categorized as corporate entities within a wider economic mode of extractive capitalism, mining value from human and natural resources. Common to all is an explicit lack of care about people's suffering and the environmental degradation caused by the corporations' activities. All have contributed in various ways to the devastating drought that has engulfed the state of California over the past decade and left brush tinderbox-dry, heightening both the number and intensity of wildfires. Many climatologists and fire behavior specialists say climate change has contributed directly to drought and extreme weather patterns. "American wildfires now burn twice as much land as they did as recently as 1970. By 2050, destruction from wildfires is expected to double again, and in some places within the United States [that is, the Southwest] the area burned could grow fivefold."[30]

A notable example of a company that has contributed to and compounded the impacts of climate change is Pacific Gas & Electric, better known as PG&E, an enormous power utility company servicing most of California. PG&E is an extractive industry in that it relies predominantly on nuclear, large hydro, and natural gas for its energy sources. It also extracts profits from individuals who have little if any choice about their energy provider, all the while acting irresponsibly toward mitigating fire risks and potential injury caused by their faulty equipment.

Since the 1990s PG&E has been involved in numerous lawsuits related to its poorly maintained power lines that have ignited many wildfires, resulting in deaths, injuries, and millions of dollars in property damage. With respect to the Camp Fire of 2018, PG&E was sued and fined $3.5 million for illegally causing the blaze that burned to the ground the town of Paradise, including eighty-four counts of involuntary manslaughter. In an emotionally charged hearing, survivors of the fire and relatives of the victims told their stories of grief in the Butte County Superior Court in June 2020. Philip Binstock, whose father died in the fire, stated, "You knew what you were doing was wrong. And rather than reduce your bonuses, you allowed failed equipment and your improper inspections to kill people."[31]

Australia's Catastrophic Bushfires

By far the worst fire disaster of recent years occurred in Australia, where devastating bushfires raged from September 2019 through February 2020. In the end about 25.5 million acres were burned, a massive swath of land compared with the 1.9 million acres burned in California's Camp Fire in 2018 and the 4.4 million acres burned in California's 2020 fire season. Most of Australia's bushfires occurred in temperate bushlands in New South Wales and Victoria. In addition to decimated forests, smoke and fire killed more than 450 people and an estimated one billion animals, including many endangered species such as the iconic koala (Figure 3). Scientists estimate it will take decades for biodiversity to recover, if ever.

The reasons for Australia's catastrophic bushfires are the cumulative impacts of human-driven climate change. Prime Minister Scott Morrison was condemned by the public and media for lax environmental protections favoring mining and the energy sector, as well as for his strategies of climate-science denial and disinformation. It was widely pointed out that he had blatantly disregarded advice from earlier governmental inquiries into fighting climate change to reduce the risk of bushfires. Morrison was also loudly denounced for bringing in police to stop public protests of the International Mining and Resources Conference in late October 2019, displaying what many commentators noted was an "authoritarian streak." As the bushfires raged across the country into 2020, they created enormous debate in

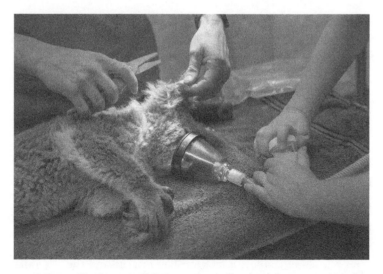

FIGURE 3. Koala rescued from an Australian bushfire, 2020. iStock.

Australia about how to deal with the social and economic impacts of such immense devastation. For many people, life will never be the same. In the words of Lynette Wallworth, a leading Australian filmmaker, we are facing "a new reality, a burning Australia. What was feared and what was warned is no longer in our future, a topic for debate—it is here." Adds professor Robyn Eckersley, an Australian political theorist, "That means massive changes in what we do and the rhythm of our work and play."[32]

Devastating fires in Brazil, the US, and Australia highlight the degree to which far-right politicians are collaborating with big businesses, presenting in turn a new set of political conditions that challenge resistance to climate change by ordinary, and increasingly disempowered, citizens. Bushfires graphically bring into focus these political conditions and the underlying antidemocratic, antiegalitarian, and racist values that sustain them.

The Nature of Fire

There have always been wildfires. For millennia bushfires have swept across low woodlands, scrublands, and grasslands, burning dry vegetation. These fires are common in places that are hot and dry, including parts of Africa, around the Mediterranean, Australia, and the southern and western regions of the United States. These fires can start naturally from a lightning strike that sets trees alight and then burn across forests and bushlands. They can also be started by human carelessness or from deliberate acts of fire management that include burning off dry vegetation as a preventative measure. Whatever the cause, over thousands of years many plants and insects have adapted to periodic bushfires, some requiring low-intensity burns to reproduce and survive. For instance, in Australia the bottlebrush tree is dependent on fire to release seeds from its specially hardened seedpods, and the fire beetle has adapted its nerve endings to help it seek out burned logs in which to lay its larvae. Another example is the amazing cogongrass, which grows in tropical areas in Asia, Africa, and Australia, as well as in the southern US. This grass is highly inflammable and actively promotes fires that burn competing plants and open land for its rhizomes to spread. As much as bushfires cause devastation, especially in rainforests where the vegetation hasn't evolved to cope with flames, fires can also provide opportunities for rejuvenation and regrowth. Fires burn decaying leaf litter, clearing land for new plants and trees to grow, and creating new habitats in which birds and animals flourish. In short, bushfires are an essential part of the natural cycle of healthy ecosystems.[33]

Many Indigenous peoples of the world know about the rejuvenating value of bushfires. With practices based on thousands of years of experience, Indigenous peoples are often experts on how to produce controlled small fires that burn overgrown vegetation and prevent huge, out-of-control bushfires from happening in the first place. Such land management requires extensive knowledge of local conditions, depending on such things as the time of year, the time of day, the degree of dryness, the type of plants and trees, and the anticipated impact on a wide range of wildlife. Noel Butler, who with his wife, Trish, used to run the Nuragunya Aboriginal culture and education camp in outback

Australia, says, "Fire in this place is our friend. Fire has been used to maintain, to look after this whole continent forever."[34] Relatively few non-Indigenous people appreciate the expert knowledge of Australia's First Peoples and their long histories of land stewardship. For instance, historian Bill Gammage, author of *The Biggest Estate on Earth: How Aborigines Made Australia*, states: "Skills like that, they have but we don't know. . . . Where the Aboriginal people are in charge, they're not having big fires. In the south, where white people are in charge, we are having the problems."[35]

Although firefighters today do use preventative burning techniques, these are not the same as the carefully controlled small fires (such as cool-fire burning) practiced by the original inhabitants of the land.[36] Firefighters' local knowledge is not adequate, and government funds for developing the required expertise are often not available. As a result, there is no real capacity for dealing with the magnitude of today's bushfires in Australia. According to Alexis Wright, who is a member of the Waanyi nation and a professor of Australian literature at the University of Melbourne:

> Aboriginal people in this country firmly believe that we are the longest-surviving culture in the world. We were raised with the knowledge that our ancestors have adapted to changing climatic conditions here for millenniums. . . . And yet our knowledge of caring for the land is questioned or largely ignored. In the face of catastrophic fires, Australia's leaders need to recognize the depth and value of Aboriginal knowledge and incorporate our skills in hazard management.[37]

A similar lack of listening to and learning from Indigenous communities about fire management is evident in many other parts of the world. Notes one commentator, "In Australia—as well as in the Amazon rainforest, the Congo, and Northern California—blatant disregard for Indigenous land management practices has contributed to the burning of the forests, the destruction of Indigenous culture, and the deaths of many people and animals."[38] Ignoring Indigenous knowledge about fire management is a global phenomenon, often leading to policies that suppress the outbreak of fire at all costs and to the buildup over decades of dry brush that, once ignited, is very hard to control.[39]

The risks of death by fire are heightened as people build homes and buildings further into the bush and communities develop on the edge of forests and woodlands. Often residents have little choice to live in such areas, taking advantage of cheap land prices, low rents, and access to mobile home parks. Firefighters' access to these increasingly remote and vulnerable communities is greatly reduced. This was the case with California's deadliest wildfire on record, the Camp Fire of 2018. But not all bushfires are in remote areas, as evidenced by the devastating Thomas Fire of 2017, which burned into residential areas in Santa Barbara and Ventura Counties in Southern California, including the very wealthy suburb of Montecito. Fires are a constant threat throughout the densely populated Los Angeles basin, frequently closing freeways and creating massive traffic congestion on already overcrowded transport networks. More recently in Australia, bushfires threatened remote rural areas as well as densely populated suburbs on the outskirts of Sydney. Out-of-control bushfires threw up massive fireballs that jumped firebreaks, creating firestorm cells that traveled ahead of the blaze and were impossible to stop, even with planes dumping fire retardant from the sky.

The point is that even with rules governing where (and how) homes can be built, and careful preventative measures including firebreaks and vehicle access routes, bushfires will continue to be more frequent, intense, and disruptive, given the conditions produced by climate change.[40] Widespread power outages, mass evacuations, daytime darkness caused by thick black smoke, untold damage to property and businesses, widespread respiratory distress requiring hospitalization, and many lost lives will be what California's former governor Jerry Brown calls "the new normal." Simply put, there is a limit to how effectively bushfires can be managed and contained. Adds Governor Brown: "This is serious . . . but this is only the beginning. This is only a taste of the horror and the terror that will occur in decades. And it will occur in various spots: in America, in Africa, in Canada. . . . It's unpredictable, other than the fact that it will get worse."[41]

Fire and Climate Change

Although there have always been bushfires, today's fires are made significantly worse by climate change. Climate change does not cause fires but creates the conditions in which fires are more intense in size and number. No longer confined to the typical fire season of late summer, bushfires now burn yearlong, making planning and management incredibly difficult. This is primarily due to the earth experiencing greater and longer periods of heat and drought, which dries out vegetation and creates a highly combustible landscape. This is clearly happening in regions around the equator that historically have higher annual temperatures. However, we are now seeing this heating pattern in cooler zones as well. Alarmingly, communities in Siberia and Alaska experienced the hottest temperatures ever recorded north of the Arctic Circle in 2020. And scientists from the National Oceanic and Atmosphere Administration tell us that July 2021 was the earth's hottest month in recorded history.[42]

This means that we will continue to experience an increasing amount of catastrophic wildfires in the foreseeable future resulting from climate instability.[43] This instability manifests on many fronts but is most obviously felt in the increasing weather extremes of droughts and floods. The swing between drought and flood creates all sorts of problems in terms of adequate water storage in dry years and potential dam and reservoir overflows in wet years. But in terms of fire management and control, it is disastrous. Massive rainfalls promote fires by encouraging new

vegetation and growth of grasslands that are highly combustible. And after fires have come and gone, floods often intensify
their devastation by creating enormous mudslides of burnt earth.
These rivers of mud, rock, and trees can dislodge buildings and
drown entire communities in meters of oozing ash and debris.

Many scientists argue that even if all intentional fire-setting
activities were completely stopped, we would still be facing a future of more and more catastrophic bushfires. This is because
global warming has reached a critical level, drastically altering
the planet's atmosphere and making its landscapes and forested
regions hotter, drier, and ultimately more flammable. For instance, in 2019 and 2020 massive fires burned across Siberia under record-breaking heat conditions. And in Europe, rising temperatures meant that in 2019 "wildfires raged as far north as
Sweden. . . . Britain had more wildfires than ever before on record." Spain too saw one of the sharpest increases in the number of forest fires, leading the European Union to declare fires "a
serious and increasing threat."[44] In August 2021, under an unprecedented heatwave, massive fires broke out in Greece north
of Athens and incinerated large sections of the island of Evia. In
the United States the Southwest is experiencing a megadrought
that makes it one of the worst droughts in the region in the past
1,200 years. Together these examples reinforce what scientific experts predict—that by 2070, a third of the world is likely to live
in extreme heat conditions. Other scientists associated with Future Earth, a global network of researchers, confirm in their report, *Our Future on Earth 2020*, "that climate change—in particular increasing temperatures—could become a prevailing force
determining fire activity in the coming decades."[45]

Wildfires are natural phenomena, but today's rapid climate
change is caused by humans. Climate change didn't just happen
as part of a natural process; it is a direct consequence of some
people—primarily white Euro-American men—abusing and exploiting the earth's resources over many centuries. It is not the

result of some divine intervention by a god punishing us for our sins. Nor is climate change a hoax or a conspiracy, as some politicians and businesspeople may tell it. There is no getting around the fact that climate change is the result of certain people's willful disregard to take care of the delicate ecological systems that we share with animals, birds, reptiles, fish, and trees.

Today we live in an era that people are calling the *Anthropocene* or the age of humans.[46] This term refers to a geological period in which human activity has been the dominant influence on the environment, first brought to public attention by Paul Crutzen in an article in *Nature* in 2002. Today there is widespread scientific consensus that we have entered a new era marked by human intervention that is changing our climate and earth systems. Jessica Weir, a scholar of human-environment relations, describes this new era: "The Anthropocene throws us a particular challenge to acknowledge those ecological connections that sustain our existence. We live within networks, webs, and other relationships with non-human (or more-than-human) others, including plants, animals, rivers, and soils. We rely on each other for food and fresh water. We are co-participants in what is happening and what will happen next."[47] However, many people, particularly those living in wealthy advanced economies, are deliberately ignoring the collective negative impacts that they make on the planet on which we all live and on which our children, and their children, will depend. As climate ecologist Corey Bradshaw argues, it is hard for ordinary people to understand the scale of the problem that "is compounded by ignorance and short-term self-interest, with the pursuit of wealth and political interests stymieing the action that is crucial for survival."[48]

Fire and Free-Market Authoritarianism

Turning around the escalating climate crisis requires enormous efforts by everyone. But it is especially important that those wielding global power, such as the political leaders of wealthy carbon-emitting countries, take a leadership role. However, what we are witnessing is just the opposite. Despite the optimism surrounding the Paris Agreement that was signed by 197 countries in 2015 pledging commitments to dealing with greenhouse-gas emissions, as of this writing in 2021, very little has been achieved on a global scale. Under the Trump administration, the United States withdrew from the Paris Agreement in 2017, joining the leaders of Iran and Turkey, who refused to sign on in the first place, and aggressively rolled back fifty years of environmental protections.[49] In 2021 newly elected President Biden reversed this action, rejoining the Paris Agreement and declaring a series of executive orders designed to tackle climate change. But it still remains unclear what material difference Biden's plans will make or whether they will have long-term impact, given the possibility of the Democrats holding on to office for only a four-year term.

Other big carbon-emitting countries have not been as open as the Trump administration about their lack of support of the Paris Agreement, with leaders simply ignoring their internationally pledged obligations. The result is that policies by far-right leaders in Brazil, Australia, and (until very recently) the United States are lining up with similar policies implemented by other radically conservative governments to effectively form a global blockade against those fighting catastrophic fires and climate change more generally. What these leaders share is a deep dependency on corporations that fund their political campaigns and help establish and maintain their political power.

I call the collusion between political governance and corporate sectors in banking, energy, agribusiness, technology, and pharmaceuticals *free-market authoritarianism*, underscoring the common antidemocratic agenda of both opportunistic political leaders and the profit logic of corporate CEOs. This collusion has been building worldwide over decades. Neoliberal policies have undermined working-class solidarities, trade unions, and the capacity of ordinary people to fight back against big businesses and the concentration of wealth and power in the hands of a few. The result is a general ideological and political swing to the extreme right. As political economists have argued, "The decline of the left has facilitated the capture and virtual elimination of democracy by neoliberalism, in order to shield market processes from political intervention and social accountability."[50]

The biggest factor in the upswing of catastrophic wildfires today is extreme right–leaning national leaders who embrace the antidemocratic priorities and economic privileges of global capitalism. These leaders are characteristically unwilling to regulate corporations, reduce greenhouse-gas emissions, plan for renewable energy production, or aggressively confront the negative impacts of climate change, which is causing hotter and drier conditions. In Australia the nation's biggest mining lobby, the Minerals Council of Australia, exerts a huge influence on domestic politics. The mining lobbyists have effectively blocked efforts to curb the country's greenhouse-gas emissions and hindered enforcement of environmental-protection legislation. In early 2020 at the height of the raging bushfires, Prime Minister Scott Morrison denied any connection between climate change and wildfires, and he came under severe attack for downplaying the role of the mining industry in contributing to the conditions for out-of-control fires. According to Adam Bandt of the Green Party, conservative leaders "bear some responsibility for what is happening at the moment because they have done everything in their power to make these kinds of catastrophic bushfires more likely."[51] In contrast to Australia, President Jair Bolsonaro's authoritarian government in Brazil doesn't deny its direct connection to devastating bushfires. Bolsonaro came to power on a political platform that openly supported burning rainforests to clear land for mining, grazing cattle, and planting soybeans. Australian and Brazilian leaders represent two responses to the problem of bushfires: angry denial of responsibility on one hand and defiant acknowledgment of collusion on the other. Whatever position antidemocratic leaders adopt, the fact is that they are collectively promoting deforestation and land clearance for agriculture and mining all over the world, and that this is in turn intensifying the scale and number of catastrophic fire events.

Against this front, an environmental justice movement is building across regions, continents, and cultures. This global movement builds on the environmental activism of the 1970s and 1980s that gave rise to green political parties in places such as Germany, Australia, New Zealand, Switzerland, Britain, and the US. But today's global environmental movement takes its cues from the global south, where the negative impacts of climate change are disproportionately experienced.[52] The movement is not hegemonic or primarily centered around political parties but rather a loose coalition of activist networks and organizations representing tens of thousands of grassroots communities with different cultural values and worldviews. A large number are linked through the Global Greens network, which seeks to implement the Global Greens Charter.[53] Uniting the broad-based movement are calls to change people's cultural values and the ways they behave and relate to each other and the natural world.[54] In various ways the movement opposes deeply ingrained capitalist logic pushing capital accumulation that necessarily involves the exploitation of natural and human resources. Increasingly, environmental groups draw inspiration from small farmers and Indigenous and nonwestern communities, arguing for the need to overcome the human/nature divide and pointing instead to a future dependent on a flourishing diversity of human and nonhuman biological life.[55] Many in the environmental justice movement appreciate that how we treat animals, forests, and oceans is profoundly connected to how we treat one another. Perhaps most significantly, green activists promote a different set of ethics from that held by most far-right leaders and the corporations and economic elites they work with and defend. This ethical grassroots perspective recognizes that societies are part of an interconnected global humanity that challenges ultranationalist policies of exceptionalism and isolationism and related political slogans such as "Make America Great Again," "Make Poland Great Again," "Make Brazil Great Again," and "Make India Great Again." Putting it simply, green activists and environmental justice groups are calling for a new political, economic, and moral framework.[56]

Thinking Through Fire

This book urges us to think *about* fires—why they happen, whom they impact, and how we can stop them. But it also urges us to think *with* and *through* fires. By this I mean that we need to move beyond treating the natural world, and by extension wildfires, as somehow different from the human world. The human/nature divide is common in most modern capitalist societies and draws on a long history separating humans from the nonhuman environments in which they live.[57] From early colonial times, the "wilderness" that Europeans encountered in the Americas, Asia and the Pacific, Africa, and the Middle East was regarded as conquerable terrain available for exploitation. Land was to be possessed, controlled, and cleared so that trees, animals, and Indigenous peoples (who were typically considered nonhuman) could be removed for houses to be built and crops to be cultivated.[58] The concept of the land being empty of all humans was necessary for colonial settlers to evoke the Doctrine of Discovery and claim legal possession. The Catholic Church issued the doctrine in 1493, and it provided the legal and moral authority for Spanish conquistadors to take possession of lands occupied by Indigenous peoples in South America. Over the centuries, the Doctrine of Discovery was adopted by other colonial powers such as the British and helped underscore a worldview in which European colonialists occupied a position of power and superiority vis-à-vis birds, animals, fish, Indigenous peoples, and enslaved Africans shipped to the New World.

From early colonial times Europeans have oscillated between treating the natural world as a resource to be used and consumed, on one hand, and defended and preserved, on the other. In both cases there is an assumption of human superiority over nonhuman species. This human/nature divide helped sustain colonial and imperial agendas for centuries, and remains very much part of western thinking today. It informs many conservation efforts and is evident in environmental movements that seek to protect an idealized nonpeopled landscape of rivers, mountains, oceans, deserts, and woodlands.[59]

Thinking *with* fire seeks to overcome the human/nature binary that posits humans as the subject and fire as the object of study. Overcoming this binary means thinking of ourselves as living within and being part of nature—what Donna Haraway calls *naturecultures*.[60] This shifts people's relationship to the environment from that of master and slave to one of coexistence and codependency, urging us to embrace "new ways of seeing, thinking, feeling, and being."[61] It acknowledges that biological organisms such as COVID-19 and the animals and insects that carry disease do not just emerge out of thin air but are the consequence of complex relations between humans and nonhumans that together make up "the web of life."[62] Putting this differently, it means that nonhuman species and ecological systems play a considerable role in shaping who we are, what we experience, whether we live or die, and how we imagine the future. Thinking *with* fire underscores that nature is not a passive resource to be pillaged or protected but constitutes a dynamic biological process of death and rejuvenation. This process is intimately connected to our material conditions and shapes in part who we are as socially conscious beings.

Thinking *through* fire opens new ways of understanding in terms of space and time that transcend our conventional nation-bound worldviews and relations with each other. Satellite images show smoke from the Brazilian fires in 2019 and 2020 blanketing South America and rising high into the atmosphere, to be seen from outer space. Similarly, satellite images show smoke from the Australian fires traveling more than nine thousand miles across the Pacific Ocean, where it was detected in cities across South America, including in Chile, Argentina, and Brazil, darkening the sky and causing ash to fall.[63] And raging wildfires across Siberia and the Arctic region in 2020 and 2021 sent smoke across the northern hemisphere into Canada and North America. Together these massive fires are causing long-term changes in rainfall patterns, soil moisture levels, ice melt flows, and oceanic currents circling the globe. These changing conditions impact agriculture and fishing food-supply chains, prompt people and animal migrations, and create new kinds of warfare over water and other natural resources—in ways we are only beginning to track and understand. (For a visual wake-up call, the Global Forest Watch Fires website provides extraordinary real-time satellite data and interactive maps that show just how extensively the planet is on fire).[64]

Thinking *through* fires as intercontinental phenomena blurring boundaries between lands, seas, and atmospheres generates different perspectives. It underscores the need to link what is happening in local communities fighting fire on the ground to global forces beyond what most of us experience or even understand. This local-global perspective helps connect what is happening in the remote Brazilian rainforests with what is happening along the heavily populated California coastline and in Australia's major cities. In connecting the dots across metropoles, regions, and continents, it is possible to highlight the similarities of crises facing diverse communities in the global south and global north. Burning jungles and coastal cities are not just parallel fire events happening in different countries but spatially interconnected and co-constitutive in terms of global formations and impacts. Wildfires challenge conventional ways of thinking that assume nation-states and national interests support the governing logic of how we must politically, socially, and spatially act in and imagine the world.[65]

Catastrophic fires also share related histories. Linking devastating fires in Brazil, the United States, and Australia are deep histories that have both caused and exacerbated climate change over centuries. These include the rise of capitalism, colonialism, industrialization, and resource extraction, as well as the more recent global escalation by far-right politicians who deny climate-change science and concurrently defund research about its economic, political, social, and cultural origins. Thinking *through* fire links the global political economy—and its relentless drive for profit—to what is happening to ordinary people as they wake up in the morning breathing thick smoke and listening fearfully for evacuation warnings. Thinking *through* fires as global phenomena with deeply interconnected historical roots highlights that we must confront intergenerational legacies of racism and exploitation. It also highlights the relatively recent evolution of state nationalism and the futility of the people of any one country thinking they can prevent the impacts of climate change by militarizing state lines or building big walls. Wildfires, floods, and rising oceans are not forces that can be stopped at the border.[66]

Perhaps most profoundly, thinking *through* fire underscores the need for different worldviews that appreciate the challenges presented by catastrophic fires that affect all societies, be they in the global south or the global north. Until very recently most people in advanced industrial societies such as the United States and Australia could ignore what was happening in developing countries with respect to climate change. The threat of rising seas drowning small island-states, for instance, was physically and emotionally removed from daily realities. But escalating out-of-control bushfires are necessarily changing this mode of complacency and disengagement. The rich and the famous in Malibu, Sydney, and Rio de Janeiro need to think not about, but together with, displaced farmers and Indigenous peoples in the Amazonian jungle and remote bushlands of California and Australia. This calls for a new kind of global consciousness that pushes back against the dominant individualism and selfish nationalism of our modern era and recognizes humanity's collective planetary future. And it concurrently calls for the valuing of much older holistic worldviews, often associated with Indigenous peoples, that underscore the deep kinship relations humans have with the nonhuman world.[67]

The following chapters paint a big picture in the hope that we can together make informed decisions about reducing human-driven climate change, which includes, among much else, scaling back devastating bushfires. **Chapter 2** focuses on the relationship between climate change and capitalism. Today many political leaders in the global north and global south are working explicitly and unashamedly on behalf of corporations, including the enormously powerful agribusiness and fossil-fuel industries. These industries are directly and indirectly producing catastrophic fire events. Relatedly, many of today's increasingly antidemocratic leaders deny climate science and the responsibility of both governments and corporations in furthering environmental degradation. **Chapter 3** discusses the global rise of ultranationalist and free-market authoritarian leaders, notably among so-called liberal democracies, who have emerged since 2010. These leaders have all been involved in using legislation to their advantage and rolling back a wide range of environmental protections that include opening national parks to mining, deforestation, and toxic dumping, as well as lowering carbon emission limits and waiving environmental regulations. Many ultranationalist governments openly ignore and tolerate—and at times even encourage—corporate irresponsibility and degradation of the environment. At the same time these antidemocratic governments are attacking the pillars of liberal democracy, such as voting rights, a free press, and public education. This makes it increasingly hard for ordinary people to assert their rights and participate in political activism that includes demanding environmental justice. In **Chapter 4** the disproportionate impacts of climate change on lower-income, racialized, marginalized, and vulnerable communities are explored. Here the concepts of structural violence and environmental racism are examined within and across countries from the global north and global south. Finally, in **Chapter 5** I discuss a future in which out-of-control fires will necessarily be the norm. Out of the blackened abyss, how can we imagine new ways of being in the world that nurtures humans and nonhumans in some sort of balanced coexistence? I conclude with a discussion of disruption and revolution in times of rising authoritarianism that points to a need to radically rethink our worldviews if we wish to turn the tide on climate change and make any measurable difference.

2

Fire as Profit

Global Corporations Rule

THE RELATIONSHIP BETWEEN climate change and the corporate pursuit of profit has a very long history. Today, however, capitalism's exploitation and extraction of natural resources has reached unprecedented industrial-scale levels. This twenty-first-century phenomenon has been enabled by the shift of power away from governments toward banking and finance sectors, where corporations call the shots. In effect, we have moved away from pursuing a liberal democratic ideal in which governments control and regulate businesses in the best interests of society, to corporations telling national leaders what to do and how to rule if they wish to remain in office. In short, the economic system is increasingly hijacking the political system under the prevailing neoliberal logic of late capitalism. According to the political theorist Wendy Brown, the "economization" of every sphere of life has desensitized us to the "bold contradiction" between a free-market economy and democratic governments that are controlled by corporate interests.[1]

In the United States, for instance, the biggest companies hold virtual monopolies in the production and distribution of goods, services, and information.[2] Corporations that come to mind are the big technology enterprises such as Amazon, Facebook, and Google; agricultural companies such as Monsanto; pharmaceutical companies such as Pfizer; banks such as JP Morgan; news organizations such as News Corp (which owns Fox News); and service providers such as AT&T and Verizon. This global power shift toward corporatism and away from democratic governance makes it very difficult for societies to control how big businesses impact a wide range of issues, including the protection of natural environments. As politics and the accumulation of capital become ever more entangled, the capacity for ordinary people to protest against environmental degradation and related injustices and to demand that governments mitigate against catastrophic weather events—such as wildfires—is rapidly diminishing.

In what follows, I look at the shift of power away from democratically based state institutions toward corporations in the US, Australia, and Brazil, and I discuss how this shift is directly linked to massive wildfire devastation from 2018 through 2020. The first case involves PG&E, an investor-owned public utility company in California that for decades has been linked to causing wildfires, including the state's most deadly wildfire in 2018. PG&E was found guilty of eighty-four counts of involuntary manslaughter for its negligence in maintaining electrical equipment and flagrant disregard for the safety of people and property. Many people were angered by the state government's failure to properly regulate PG&E's activities, including implementing protocols on safety and risk assessment.[3]

I then turn to Australia and Brazil and their respective mining and agribusiness industries. In both instances the privatization and subsequent massive exploitation of lands and natural resources have resulted in catastrophic wildfires. Today's national leaders of Australia and Brazil work aggressively to defend corporate interests and in practice are complicit in the burning of forests and destruction of unique ecological systems and cultural heritage in the name of profit.

Our capitalist system depends at its core on making money from the exploitation and commodification of human and nonhuman resources. Some analysts call this *extractive capitalism* for its carefully engineered extraction of financial value from people and things. To better understand the California, Australia, and Brazil case studies, I first take a detour to discuss how capitalism—and specifically the late capitalist practices of the twenty-first century—has enabled a worldview that sees the burning up of our collective futures as justifiable for the financial gain of an increasingly select few.

Capitalism and Extractivism

The march of modern capitalism has been accompanied by an extraordinary debasement of human and nonhuman environments. This process is often called *extractivism*, referring to the removal of resources, including land and minerals, and turning them into commodities to sell for profits that benefit the extractors and not the local people whose lands and resources have been taken. It involves a relation of power between the extractor and what is extracted that is not reciprocated and is fundamentally exploitative. We typically think of this process first appearing with eighteenth- and nineteenth-century industrialization, linked to such things as the deep coal mines of northern England or the Dutch and Norwegian commercial whaling fleets of the open seas. However, it is a process that has been going on since the Italian Cristoforo Colombo (widely known as Christopher Columbus) landed in the West Indies in 1492, enabling the Spanish to set up a colonial empire using the Caribbean islands as their military base and send conquistadors into Central and South America. Under the legal and moral authority of the Doctrine of Discovery, the Spanish defeated the Aztec and Inca peoples and conquered an enormous part of the Americas, extracting in the process gold, crops, knowledge, and slaves.[4] A related process was that of the English invading Ireland in the fourteenth and fifteenth centuries and setting up one of the earliest systems of colonial dispossession, marking the transition from feudalism to capitalism.

From its earliest beginnings, capitalism as an economic enterprise involved the exploitation of both human and nonhuman resources. Human resources primarily involved the use of unpaid labor, with people forced to join armies, farm sugar and tobacco plantations, work on docks and boats, clean houses, cook, care for children, and perform a full range of domestic duties. But it also involved more horrifying and insidious forms of extractivism in the form of colonialists using enslaved women's bodies to give birth to new generations of children to sell as property. Similarly, colonialists used enslaved people's minds to extract local knowledge about all sorts of things, such as how to find water, where local rivers ran, what animals threatened people and crops, and how best to survive in what for many Europeans were strange new lands.

Extractivism has been intertwined with capitalism in its various forms over the centuries, but since the 1970s it has become globally triumphant under neoliberal economic theories of unlimited resources and growth.[5] For the past five decades, the market-driven logic of late capitalism, coupled with a global development infrastructure implemented through the World Bank and International Monetary Fund, enabled the relentless extraction of cheap labor and natural resources by transnational corporations and the western governments that supported them. Mercenary armies were often hired to carry out these activities. This neocolonial looting of resources took place predominantly in developing economies throughout Africa, the Middle East, Asia, and Central and South America in the later decades of the twentieth century.[6] It left in its wake massive degradation of societies and natural environments, with little compensation given to local communities destroyed for generations by invading opportunistic industries (Figure 4).

FIGURE 4. Judy Seidman, South Africa, 2020. "Our world belongs to all who live on it" is based on the statement of South Africa's 1956 Freedom Charter (the founding document of the antiapartheid struggle): "South Africa belongs to all who live in it."

Importantly, this strategy of extractivism was pushed hard by mainstream Euro-American neoliberal economists influenced by the teachings of people such as Friedrich A. Hayek, Milton Friedman, and James M. Buchanan, each of whom received the Nobel Prize for economics in 1974, 1976, and 1986, respectively. As analysts have noted, "With few exceptions, economics as a discipline has been dominated by a perception of living in an unlimited world, where resource and pollution problems in one area were solved by moving resources or people to other parts."[7]

Today the opportunistic looting of the global south has spread unabashedly into the global north as well. Neoliberal logic, coupled with increasingly antidemocratic governments, has enabled the relentless mining and exploiting of cheap labor and natural resources in advanced economies such as the United States, Britain, Australia, Italy, Spain, and so on. We see this in the ballooning numbers of impoverished and precarious workers forced to hold multiple low-paying and insecure jobs, giving rise to extraordinary socioeconomic inequalities within and across all countries of the world.[8] And we see this in the global land-grabbing rush as transnational corporations and governments buy or lease large-scale landholdings for mining, agricultural production, and access to water.[9]

Today's massive land grabbing has taken on a different profile from earlier eras.[10] The 2008 economic recession revealed deep issues of food shortage, including limited access to water, in many parts of the world. In a sense a global panic emerged as countries contemplated declining capacities to feed their populations. Food security became a heightened concern among wealthy nations that up to that point had associated it with the impoverished regions of the global south. Concerns over food and water supply have led to the biofuel and agribusiness industries aggressively appropriating access to lands. According to GRAIN, a Barcelona-based NGO that works on issues of food sovereignty, the 2008 global financial crisis and world food crisis helped spawn "a new and disturbing trend toward buying up land for outsourced food production."[11] Countries such as China, Korea, Japan, and the Gulf States have bought huge swaths of land in places such as Burma, Sudan, Pakistan, Brazil, and Uganda. Also joining the global land grab are corporations and government agencies based in the European Union, Britain, the United States, and Canada.[12] Facilitating these predatory governments and corporations is a network of investment banks, such as Deutsche Bank and Goldman Sachs, which help with agricultural investment funds, land leases and purchases, and participation in special economic zones.

According to the sociologist Saskia Sassen, land grabbing is the hallmark of today's global capitalism and has enormously exacerbated global inequality. She writes that in contrast to earlier periods of capitalism, "the natural resources of much of Africa, Latin America, and central Asia are more important than the people on those lands as workers or consumers."[13] So it's not surprising that Bill Gates, one of the five richest people in the world, has in recent years become the largest private owner of farmland in the United States. Comments Nick Estes, a scholar and a citizen of the Lower Brule Sioux Tribe, "Land is power, land is wealth, and, more importantly, land is about race and class. The relationship to land—who owns it, who works it, and who cares for it—reflects obscene levels of inequality and legacies of colonialism and white supremacy in the United States, and also the world."[14]

The global land grab confirms what most of us already know: that the planet's resources are not infinite, and extractive capitalism is not sustainable. Moreover, the global land grab graphically reveals the negative impacts of extracting value from people, animals, and lands without understanding long-term impacts or providing for local people suffering from its detrimental consequences. Appreciating the limits of extractive capitalism is not new, and open criticism of this exploitative system has been around since the 1960s. One of the earliest and most acclaimed critics was Rachel Carson, who in her book *Silent Spring* (1962) focused on the negative consequences of spraying DDT on fragile ecosystems. Carson publicly condemned the careless use of toxins by pharmaceutical and agricultural industries, gaining widespread popular support for her outspoken activism. She became a hero of the emerging environmental movement that witnessed the Santa Barbara oil spill in 1969 and established Earth Day in 1970, when 20 million people in the United States took to the streets to demonstrate against the exploitative power of corporations.[15]

Second among the major critics were the authors of the MIT report, led by Donella H. Meadows, titled *The Limits to Growth* (1972), which was commissioned by the multilateral Club of Rome. In contrast to Carson's popular audience, the report spoke to scientists using computer modeling to predict global limits to industry, food production, and sustainable population. Both *Silent Spring* and *The Limits to Growth* were and remain environmental best sellers around the world. But at the time of their publication and in the ensuing years, both books were widely rejected by businesses and economists, and were the targets of publicity campaigns that denounced the scientific evidence of their arguments.[16] More recently, reevaluations of *The Limits to Growth* confirm the report's basic premise that if nothing is done to halt the exploitation of resources, the planet faces a massive decline in biodiversity by the latter part of the twenty-first century. Unfortunately, this prediction is already very evident.[17] In a reassessment of the report, Christian Parenti has argued that it was "a scientifically rigorous and credible warning that was actively rejected by the intellectual watchdogs of powerful economic interests. A similar story is playing out now around climate science."[18] Adds Jason Hickel in his latest book *Less Is More: How Degrowth Will Save the World*, "in a growth-oriented economy, efficiency improvements that *could* help us reduce our impact [on the planet] are harnessed to advance the objectives of growth—to pull ever larger swaths of nature into circuits of extraction and production. It's not our technology that's the problem. It's growth."[19]

Today, extractive industries remain the driver of the global political economy, and gross domestic product (GDP) figures, stock value, and annual profits are the measure of their success. This is despite the early warnings by Carson and other critics and activists, as well as a global scientific consensus affirming that the planet has finite resources. Writes Vandana Shiva, the internationally recognized peace activist, "The global corporate economy based on the idea of limitless growth has become a permanent war economy against the planet and people."[20]

To understand the nature of this war and the continuing dominance of our shortsighted destructive economy, the public intellectual Naomi Klein offers some insight. She argues that the resource-depleting economic model requires a particular worldview that allows some people to treat the earth with violence: "to dig and drill out the substances we desired while thinking little of the trash left behind, whether in the land and water where the extraction takes place or the atmosphere, once the extracted material is burned."[21] Klein goes on: this mindset also extends to how some people treat others with violence, reducing individuals "either into labor to be brutally extracted, pushed beyond limits, or, alternatively, into social burden, problems to be locked out at borders and locked away in prisons or reservations."[22] Related to this are what are sometimes called "sacrifice zones," where extractors consider some places, and the people that occupy them, sufficiently worthless that they can be poisoned, burned, leveled, and destroyed.[23] Violence—toward people and environments—is very evident in the way companies, and the governments that increasingly support them, relentlessly pursue industrial-scale extraction and profit in the twenty-first century.

The next section explores the blatant disregard for human and nonhuman life by profit-driven corporations in the US, Australia, and Brazil that predictably led to the catastrophic wildfires of recent years.

California and PG&E

Pacific Gas and Electric Company (PG&E) is a US investor-owned utility that provides natural gas and electricity to much of Northern California and supplies more than five million households. It is one of six electric utilities in California, but it is a corporate giant and in effect operates as a state-regulated monopoly. PG&E started off as a gasworks company in San Francisco during the gold rush in the mid-nineteenth century. The company quickly built up a thriving business in the burgeoning city, where the demand for lighting in warehouses, shops, and shipbuilding industries was high. PG&E was always an innovative and entrepreneurial company, building the longest pipeline in the world in 1930 connecting San Francisco to the Texas gas fields. It also purchased competing gas and utility companies and secured access to water necessary for producing hydroelectricity throughout the 1920s and 1930s. One of its subsidiary projects was establishing the Sacramento Street City Railway, a fleet of seventy-five streetcars servicing downtown San Francisco powered by its hydroelectric plant. By 1935, PG&E had consolidated a huge energy monopoly that served all of Northern and Central California.[24]

In the years after World War II, PG&E expanded its business in natural gas production and started producing nuclear power at the Vallecitos Nuclear Center in Northern California in 1957. This was the first privately owned and operated nuclear power plant providing electricity to a public utility grid. PG&E then began construction of another major nuclear facility in 1968, the Diablo Canyon Power Plant. The plant took years to build and be approved by environmental agencies, but it finally went into full production in 1985. Today, PG&E is aggressively trying to dismantle the plant because natural gas and hydroelectricity are much more financially lucrative investments.[25]

PG&E has long been associated with causing wildfires across California. In 1994 it caused the Traumer Fire, for which it was found guilty of 739 counts of criminal negligence. In 1996 the company was held legally liable for another fire that broke out in downtown San Francisco. Then in 1999 PG&E was found responsible for poor management of dry vegetation around power lines that caused the Pendola Fire, which burned twelve thousand acres of woodlands in the Plumas and Tahoe National Forests. The fines and penalties for the fires, coupled with the huge costs PG&E suffered when it was forced to buy electricity from out-of-state suppliers at inflated prices, forced the utility to file for bankruptcy in 2001. Since then PG&E has been involved in case after case of negligence and liability. In 2010 a huge gas explosion occurred in San Bruno, California, when an old gas pipe blew, killing eight people and setting fire to one hundred homes. In 2014 an inquiry found that PG&E "knowingly and willfully" disregarded the laws surrounding gas pipeline safety. In 2015 PG&E was held responsible for the Butte Fire, which burned 70,868 acres, killed two people, and destroyed hundreds of homes in Central California. Throughout 2017 PG&E was again held liable for thirteen separate fires that burned across Northern California, causing death and devastation. The fires ignited due to the company's poor equipment and maintenance of power lines, as well as its failure to implement required safety standards to cut back brush and vegetation. But it was the Camp Fire of 2018, then the second-largest and most deadly fire in California's history, that caused massive public outrage. The fire started when a faulty electrical transmission line broke free from a nearly hundred-year-old tower and rapidly spread into rural and urban areas, killing many people, burning to the ground the town of Paradise, and destroying more than eighteen thousand homes and structures.[26]

PG&E was sued for starting the Camp Fire and charged with 84 counts of involuntary manslaughter. After hearing testimony from survivors of those killed, Judge Michael Deems of Butte County ordered the company to pay $3.5 million in damages. To many of the survivors, the fine seemed totally inadequate, given that the CEO, Bill Johnson, had an annual salary of $2.5 million (plus stock awards worth at least $3.5 million a year), and net assets of the company were estimated to be worth $85.2 billion. But apart from money, many survivors were distressed by the decades-long pattern of company negligence and failure to abide by safety provisions. Most of all, people were angered by the company's explicit indifference to loss of life and property, and the state's complicity in failing to ensure PG&E act legally. According to Joseph Downer, whose brother Andrew died in the fire, "I don't believe justice is served by a $3.5 million fine. If ever there was a corporation that deserved to go to prison, it's PG&E." Tommy Wehe, whose mother burned to death in her truck as she tried to flee, argued that the acceptance of guilt by PG&E didn't amount to anything if the company failed to change its practices and prevent future deaths. Judge Deems indicated that he too was frustrated by the inadequacy of the fine, saying, "If these crimes were attributed to an actual human person rather than a corporation, the anticipated sentence . . . would be ninety years to be served in state prison."[27]

PG&E and Climate-Change Denial

In 2019 PG&E filed for bankruptcy a second time. The case attracted widespread attention among policy, planning, and insurance sectors concerned about financial risks to utility companies, given increasing droughts and catastrophic wildfires. PG&E is seen by some economists at the *Wall Street Journal* to be "the first major corporate casualty of climate change."[28] Faced with $30 billion in liabilities for fire damage, and facing 750 additional lawsuits for compensation, PG&E was keen to shield itself through the legal protections provided by corporate bankruptcy laws.

There is some irony in PG&E being the first case of climate-change bankruptcy, given the company's long history of denying climate change even existed. From 1989 through 1992 the company was a member of the Global Climate Coalition, which led media campaigns intended to stop the regulation of greenhouse-gas pollution. At the time PG&E was fully aware that its coal-burning plants were contributing to global warming. It became involved with a broad spectrum of energy providers to sow disinformation and skepticism about climate change to ensure continued deregulation and maximum profit margins.[29] Notably, PG&E was not alone in its insistent denial of climate-change science.[30] For instance, as early as 1981 an expert with Exxon admitted in an internal memo to his colleagues that carbon dioxide emissions were linked to global warming. Despite that knowledge, Exxon participated in a widespread climate-science denial campaign for years and spent more than $30 million on lobbyists, think tanks, and researchers to promote climate-science skepticism.

This widespread disinformation campaign was a direct response to mounting scientific evidence that climate change was caused by greenhouse-gas emissions. In 1988 the US Congress and the general population were put on notice that greenhouse gases were building and effectively warming the atmosphere by the climatologist James Hansen.[31] This was internationally confirmed by the founding of the Intergovernmental Panel on Climate Change (IPCC) that same year. From the late 1980s and into the 1990s, an organized disinformation campaign went into full swing to create confusion around climate-science findings and attack the legitimacy of climate scientists. Although the campaign started in the United States, by the mid-1990s it had filtered out to other countries such as Britain, Australia, and Canada. Conservative politicians and industry networks heavily promoted nonacademic testimony as expert evidence, arguing there was no scientific basis in climate-change claims.[32] Moreover, the small number of atmospheric scientists openly skeptical of climate change received compensation from oil and gas companies. For instance, Willie Soon from the Harvard-Smithsonian Center for Astrophysics was paid more than $1 million by ExxonMobil, the American

Petroleum Institute, and Koch Industries to dispute scientific evidence about the impact of human-generated greenhouse gases.[33] According to the French philosopher Bruno Latour, with mounting evidence on climate change and a finite limit to the planet's resources, political and economic elites had to come to terms with a new reality that threatened their power and profits. Hence company executives, such as those at ExxonMobil, had to go into a "frenetic campaign to proclaim the non-existence of the threat," and *"stop pretending, even in their dreams, to share the earth with the rest of the world."*[34]

One of the reasons for the disinformation campaign's success was the ability of corporations to control the public narrative. Professional skeptics financed by conservative think tanks were given overblown coverage in mainstream mass media. Editors and journalists, particularly those working for Fox News and other news outlets owned by media mogul Rupert Murdoch, were often pressured into covering certain stories.[35] The result was that "reporting on climate in the United States became *biased* toward skeptics and deniers."[36] This bias was also evident in Australia, where Murdoch's company, News Corp, controlled many news outlets. Today in Australia, News Corp owns about two-thirds of daily newspaper circulation through its 140 newspapers and dominates the print, online, and broadcast media landscape. The cumulative result is that for decades climate science has been denounced by conservative politicians, supported by corporations and economists pushing a profit-driven neoliberal agenda, and reinforced through extreme-right global print and social media companies such as Murdoch's News Corp.

As fires burned across Australia in early 2020, James Murdoch, Rupert Murdoch's younger son, publicly rebuked his father's news empire for denying climate change and the clear evidence connecting it to raging bushfires. In October 2020 former opposition prime minister Kevin Rudd posted a petition asking the government to establish an inquiry into the dominance of

Australian media by News Corp. Rudd called Murdoch's media empire a "cancer on our democracy" for its long history of undermining efforts to combat climate change, trying to interfere in elections, and pushing racist rhetoric and extreme-right policies on immigration.[37] Clearly, in the context of corporate-owned media, the ability to control the narrative on climate change and other scientifically proven phenomena—such as the COVID-19 pandemic—continues to be highly politicized. This is the case in Australia, the United States, Canada, Britain, and many other countries as well. Given the lack of regulatory oversight of the corporate giants of news and social media, control over the public narrative will most probably remain highly politicized and biased into the foreseeable future.

PG&E will survive its current bankruptcy filing, just as it survived after its 2001 bankruptcy. Its stock values may fluctuate, but inevitably it will continue to flourish. As essentially a state-regulated monopoly, it is "too big to fail." State legislators have limited choice, given the costs involved and the state government's financial dependence on PG&E's ability to raise money in the stock and bond markets. For its part, PG&E has promised to trim and cut trees along its power lines and do better equipment maintenance (a promise it has made and broken in the past). More drastically, PG&E has started turning off its electrical grid when it gets hot and windy to stop downed trees and sparks igniting, sending hundreds of thousands of people into darkness. This means that Californians are severely challenged, having to put up with having no power for days to avoid endangering people, houses, and animals with massive fire disasters. And all the while PG&E continues to make huge profits and exerts enormous political and economic influence within California's governance.[38] This behavior is similar to many other private-sector companies that expect—like Bank of America, Citigroup, Goldman Sachs, and other big banks that were bailed out in the wake of the 2008 recession—to operate without robust state regulation and then demand that the state bail them out when faced with economic collapse.

Australia's Quarry Economy

In Australia, as in California, big corporations exert huge political and economic power over the running of the country. One of the most dominant corporate sectors in Australia is fossil-fuel production and the mining industry. In many ways, mining epitomizes extractive capitalism, with its brutal digging, scraping, and squeezing value out of the ground, unnaturally upturning and exposing minerals and precious metals from deep within the planet.

Although the mining industry has had its fluctuations in terms of profitability, particularly with respect to a global trend toward renewable energy, it remains an incredibly important element of Australia's economy. Since the 1990s Australia has become overwhelmingly a "quarry economy" and is home to some of the world's biggest and most ambitious mining projects. Australia plays an extraordinarily large role in terms of the world's production and export of natural minerals and gas (exported as liquefied natural gas, or LNG). For instance, in 2019 Australia was the world's largest producer of iron ore and bauxite, and the second biggest producer after China of gold. It was also responsible for 15 percent of the world's output of aluminum, 12 percent of uranium, and 5 percent of silver. In terms of exports, minerals make up approximately 35 percent of the country's overseas trade. Australia is the biggest exporter of coal, with much going to China and India. It is also the second biggest exporter of gold and uranium. In sum, Australia's economy is highly dependent on selling nonrenewable fossil fuels and minerals for its growth and prosperity.[39]

Mining has a long history in Australia. Throughout the nineteenth century, mining, and specifically two major gold rushes in the 1850s and 1890s, helped shape the new British colony. The promise of fortune lured waves of immigrants from around the world to the continent. Gold rushes and subsequent expansion of the mining industry shifted the country's demographics as well as its long-term economic orientation. Today, Australian society hotly debates the controversial politics of mining and extractive industries, and many Australians are concerned about the country's contribution to climate change. Mining also causes pollution, flora and fauna destruction, drought, and extreme weather events. Around the country protests have erupted in opposition to the powerful mining lobby, the Minerals Council of Australia, and its influence on domestic and international policies and the country's lack of adequate environmental protection legislation.

Kalgoorlie's Super Pit

One of the most spectacular and horrifying examples of a massive mining project is the Super Pit gold mine. The Super Pit is located on Kalgoorlie's Golden Mile, a stretch of land known for its condensed and deep gold deposits, situated about six hundred kilometers (about 370 miles) east of Perth in Western Australia. In 1893 prospectors first discovered gold in the area. Since that time countless miners have created more than thirty-five hundred kilometers (approx. 21,000 miles) of underground tunnels and shafts that extend deep into the ground to about six hundred meters (approx. 2,000 feet). In the late 1980s, a century after gold was first discovered, the businessman Alan Bond bought up all the mining leases in the area with the plan to dig a huge single pit. However, faced by financial difficulties, Bond formed a company called Kalgoorlie Consolidated Gold Mines that bought up all the leases and developed a huge mine that is approximately 3.7 kilometers long by 1.55 kilometers wide and more than 660 meters deep (more than two miles long, nearly a mile wide, and 2,500 feet deep).

The Super Pit is breathtaking in its scale and impact on the landscape, dwarfing the adjacent town and residential streets of

Kalgoorlie, as well as the thirty thousand inhabitants who live on the pit's parameter. The Super Pit can be seen from outer space. Around 15 million metric tons of rock are moved from the Super Pit each year, loaded out on huge trucks that often operate twenty-four hours a day. Each truck is the size of a small building and uses approximately $8 million of fuel and $3 million worth of tires to run. Tours of the Super Pit and a viewing platform attract a certain crowd, drawn to the terrible might of humankind's impact on the earth. According to one excited commentator, "Next to the town of Kalgoorlie sits a gaping maw, wider and deeper than many canyons and worth millions of dollars . . . a *day*. It's called the Super Pit, the largest open-pit gold mine in Australia, and one of the largest in the world."[40]

The environmental impact of the Super Pit is an ongoing and hotly debated issue. According to one report put out by the Government of Western Australia titled *A Future Beyond the Pit*, there is an urgent need to acknowledge the devastation mining has had on local flora and fauna, as well as its damage to the ancestral lands and contemporary lives of Indigenous peoples, many having unique cultural and spiritual relations to the land. According to the report, the Great Western Woodlands is an environment of "global significance."[41] The Woodlands is home to a range of animals and vegetation that face extinction due to bushfires and climate change, and without proper management the critical habitat will be destroyed forever. Apparently, there are plans afoot to build a two-hundred-hectare park, Karlkurla Bushland Park, to encourage natural regrowth of the landscape some miles north of the town. However, despite the good intentions of some community members, the plan seems ineffectual, given that the pit spans twenty-six thousand hectares and the devastation to the physical environment extends well beyond the actual mining site itself. Other ideas about what to do with the Super Pit when the mine is projected to close in the year 2035 remain uncertain. What is clear is that the transnational mining companies involved

will not be held accountable for most of the subterranean degradation and long-term toxic impact on all Australians and their
unique flora and fauna.

Mining, Climate Change, and Catastrophic Bushfires

Although the Super Pit is the largest and most recognized of the
mining "voids," what to do with open mines in general is a big
issue in Australian society. Mining companies have typically not
been legally required to provide for rehabilitation of the pits or
to fill them back in once they close. According to a report sponsored by the Australian Conservation Foundation, "the use of
large open pits produces desolate landscapes riddled with pits,
dumps, pollution, and subsidence events." The report goes on:
"These ticking time-bombs of environmental, cultural, and social impacts will ultimately interact and accumulate and will require technical and political solutions of great complexity." In
addition to mines not yet closed, there are approximately fifty
thousand abandoned mines scattered across the country that
have been left to leach tainted water into the water table and
ooze toxins into the landscape. These environmental disasters
join those of denuded mountain tops that were logged to serve a
mining smelter in the 1900s (Queenstown, Tasmania) and more
recently a despoiled world heritage site plagued by hundreds of
spills and leaks from underground uranium mining (Kakadu National Park, Northern Territory).

Some researchers estimate that the cumulative impact of mining has left 10 million hectares of land affected by pollution and
destruction of vegetation and bushland, contributing to loss of
habitat, water tables drying up, soil erosion, drought, and extremely dry conditions.[42] The scientific evidence is very clear that
mining played both a direct and an indirect role in creating the
environmental conditions for Australia's catastrophic bushfires
of 2019 and 2020. Mining contributes significantly to climate
change, and as climate change steadily heats the vast Australian

continent, all indications are that such fires will intensify in number and scale into the future.

However, the prime minister of Australia, Scott Morrison, denies connections between mining, climate change, and devasting fires. As a pro-business ultranationalist leader, Morrison has determinedly sided with the mining sector and reaffirmed the country's economic dependency on fossil-fuel production and mineral exports. Morrison is the latest prime minister to play to the political and economic power of the mining industry and affirm its extraordinary influence over the nation's attempts to pass legislation dealing with climate change.[43] Sometimes mining entrepreneurs (what some call mining magnates) become politicians or leaders of political parties, as in the case of Clive Palmer, who is a multibillionaire and at one time owned mining corporations, golf courses, resorts, and a football club. In 2013 he created his own political party, the Palmer United Party, and after winning a seat in the federal election sat as a member of the House of Representatives for one term.[44] More typically, mining entrepreneurs exert their power through lobbyists and political networks, often hiring former politicians or having their own lobbyists appointed by government leaders.[45] According to the former opposition prime minister, Kevin Rudd, "Glencore, Rio [Tinto], and BHP ran sophisticated political operations against my government, both on climate change and the mining tax."[46] These campaigns lobbied hard against government plans to tax corporate profits and limit carbon emissions, and were in the end very successful in protecting the companies' interests on both fronts.

George Rennie, an expert on lobbying, agrees that the Australian mining industry is extremely powerful, noting, "The power of the resources sector comes from its profits—its ability to spend on donations and gifts, as well as its own political advertisements if it chooses." He goes on: "There is disproportionate power, so if you want to lobby government for something that the resources

sector does not want, you're very unlikely to get your way." Adds
Dom Rowe from Greenpeace Australia Pacific's program, "Considering the vast network of influence and direct access that fossil-
fuel executives and lobbyists have to senior government ministers,
it is little wonder that the Coalition [government] has yet to take
any meaningful action on reducing Australia's greenhouse-gas
emissions, which have been rising for over four years."[47]

Australian society, as in the United States, is bombarded with
disinformation about climate science. Historically, big businesses
and transnational corporations have worked hard to deliberately
create confusion about whether climate change is really happening and whether it is a serious risk to humankind.[48] In a 2020
survey conducted by the News and Media Research Centre at the
University of Canberra, almost one-fifth (18 percent) of Australians deny climate change or don't think it is a serious issue (the
US, Sweden, and Norway have even higher numbers!).[49] Most of
these people live in rural areas and support Morrison's extreme-
right policies. The survey points to the enormous influence of
Murdoch's media company, News Corp Australia, and its long-
term ties with the conservative Liberal Party.

Australia's culture of climate-science denial has served Prime
Minister Morrison very well. His political career depends on the
financial support of the mining lobby and far-right economic
conservatives who favor the mining sector. For instance, when
treasurer in February 2017, Morrison dramatically mocked politicians concerned with greenhouse-gas emissions by bringing into
Parliament a large lump of coal and declaring, "Don't be afraid,
don't be scared" to opposition members (Figure 5). Environmentalists around the world criticized the antics, which were considered unprofessional, dismissive, and in extremely poor taste. Social media hummed with angry tweets, such as "Day 2, 50 C,
Koalas dying. Cattle dropping dead, crops cooking in fields. Aus
Minister for Agriculture says 'burn more coal.'" Another tweet
read: "Aust'n temp records smashed. Fire danger 'catastrophic.'

FIGURE 5. Treasurer Scott Morrison hands Deputy Prime Minister
Barnaby Joyce a lump of coal during question time in the House of
Representatives at Parliament House in Canberra in a publicity stunt
supporting the mining sector, February 9, 2017 (AAP Image / Mick
Tsikas).

Govt ministers laugh as wave lump of coal at those who want
clean energy. Wicked."[50]

A few years later, in the run-up for reelection in May 2019,
Morrison campaigned hard against the progressive Labor Party,
which had pledged to take tougher action on climate change (and
minimize tax loopholes for wealthy Australians). For many it was
an election that hinged on the parties' respective policies toward
the environment. Morrison was financially backed by the mining industry and made it very clear that if he was reelected, he
would not introduce new taxes or environmental restrictions on
that sector. He also indicated that he would open new coal mines
such as the controversial Adani Carmichael mine, which deeply
threatens the Great Barrier Reef and unique Queensland biodiversity habitats. As one commentator noted, "The battle over the
[Adani] mine has been the usual sordid tale of fossil-fuel industry

development, in which a rich, powerful, politically connected corporation gets its way with weak and corrupt politicians."[51] In a surprise outcome, Morrison marginally won reelection by one seat to head a Liberal-National Party coalition. World-renowned conservationist David Attenborough could not hide his displeasure, saying of the government and its pro-mining laws, "You are the keepers of an extraordinary section of the surface of this planet, including the Barrier Reef, and what you say, what you do, really, really matters. And then you suddenly say, 'No it doesn't matter . . . it doesn't matter how much coal we burn . . . we don't give a damn what it does to the rest of the world.'"[52]

In Australia there is no mistaking the reality that mining corporations exert significant influence over the political landscape, and that this influence accounts for the country's notable lack of environmental legislation. Through its extensive mining sector, Australia contributes to escalating climate change both across its own vast continent and around the world. As a self-declared advocate of mining interests, Morrison is a puppet of aggressive corporate capitalism that acts without concern for the deaths it causes or the environments it decimates. It appears that for Morrison, out-of-control bushfires are an unfortunate but nonetheless justifiable trade-off for securing enormous corporate profits and maintaining political power. Thus, when catastrophic bushfires broke out toward the end of 2019 that would turn into the worst bushfires in the country's history, there was widespread anger directed at Australia's prime minister, who many people felt was directly responsible for the devastation. A poem addressed to him by Anne Layton-Bennett, a journalist from Swan Bay in Tasmania, sums up the country's widespread sentiment at the time:

how many trees have to burn
how many animals
birds, plants, insects have to die
how many people have to suffer
and

watch their homes, their farms, their animals
and their lives
be destroyed
before you bring yourself
and your government
to acknowledge
and recognize we're in the middle of
a crisis here
an unprecedented environmental
and fossil-fuelled Armageddon
facilitated and fostered
by you[53]

Brazil's Insatiable Agribusiness

On the other side of the world from Australia, Brazil's economy is also deeply entangled with extractive capitalism. However, in Brazil devastating fires are not the unfortunate though predictable by-product of mining as in Australia, but rather an explicit strategy of agribusiness management. As the biggest country in South America, Brazil depends heavily on industrial-scale agriculture and the enormous profits it makes from the production and global export of soybeans and beef. This booming agriculture sector requires vast expanses of farmland for growing crops and grazing cattle. Thus, it is not entirely surprising that the radical-right government led by President Jair Bolsonaro has been putting policies in place that allow the deliberate burning and clearing of rainforests.

Like Australia's prime minister, Scott Morrison, Brazil's president, Jair Bolsonaro, is a far-right pro-business leader. Like Morrison, Bolsonaro came to power on a campaign pushing an extreme free-market policy that specifically spoke to domestic big-business interests, foreign investors, and the country's flourishing system of extractive capitalism. Bolsonaro promised the agribusiness sector that if he was elected, he would dismantle environmental protections and allow the burning of vast swaths of rainforest for soybean production and grazing. This promise echoes that of Morrison to the mining sector to open controversial mines and roll back environmental legislation.

Bolsonaro was trained in the military, and his primary political base has always rested on his military connections. For decades he served in Congress, praising Brazil's earlier military dictatorship (1964–85), denouncing the idea of democracy, and openly attacking those on the political left, women, and Indigenous and LGBTQ communities. In his run for president, he knew he had to appeal beyond his military base and attract a coalition of factions including businesses, Evangelicals, and the judiciary. Paulo Guedes was carefully chosen as his campaign's economic advisor. Trained at the University of Chicago under Milton Friedman and well versed in the neoliberal values of the Chicago School, Guedes satisfied Brazil's business sector. His credentials were strong—previously teaching in Chile under the Pinochet dictatorship and back in Brazil rising to fame as a successful investment banker.[54] On January 1, 2019, Guedes took office as minister of the economy in Bolsonaro's new administration. In the subsequent months he aggressively advocated austerity initiatives that included privatizing state enterprises, deregulating a broad range of corporate checks and balances, and promoting mining and industrial agriculture.

Important to Bolsonaro's continuing leadership is support from a powerful group in Congress, the União Democrática Ruralista (UDR), known locally as the *ruralistas*. This right-wing association of landowners and farmers reflects the interests of foreign investors and specifically the multibillion-dollar mining and agribusiness sectors. These sectors have played a considerable role in Brazil's politics and economy for decades. Since the 1970s development projects have seen mines dug, rainforests logged, roads cut, dams built, and land cleared for cattle grazing and the extensive growing of coffee, sugar, maize, and other crops.[55]

However, in the 1990s rainforest deforestation received international condemnation and forced Brazil to take political action against farmers and loggers. Brazil initiated new environmental regulations that allowed Indigenous communities to govern specially designated lands (about 13 percent of Brazil's territory that includes large areas of rainforest). Also established were ecological reserves to protect the rainforest's unique biodiversity, such as Jamanxim National Forest. Brazil was hailed as a leader in environmental conservation with its effective campaign that reduced deforestation by about 80 percent from 2005 through 2012. And in 2015 Brazil committed to eliminating illegal deforestation in the Amazon and reforesting 12 million hectares when it signed the UN Paris Agreement on climate.[56]

Regionally, Brazil functions as a leader with respect to central South American rainforest management. The rainforests span nine countries, with Brazil having the majority of rainforest within its borders (60 percent), followed by Peru (13 percent), Colombia (10 percent), and smaller amounts in Venezuela, Ecuador, Bolivia, Guyana, Suriname, and French Guiana. According to Lisa Viscidi and Nate Graham of the Inter-American Dialogue's Energy, Climate Change, and Extractive Industries Program, "The country's efforts in this respect matter on a global scale: The Amazon is estimated to contain 10 percent of the world's biomass, absorbing and storing massive amounts of atmospheric carbon dioxide."[57]

Soybeans and Deforestation

Despite Brazil's history of regional and global leadership on defending rainforests, legal protections of the Brazilian rainforest have been steadily undermined in recent years. This is in large part due the increasing economic importance of soybean production across South America and particularly in Brazil since the late 1990s. Soybean is an extremely flexible crop and when processed becomes biofuel, soybean oils, and soybean oil meal to feed cattle, pigs, and other livestock.[58] The global market value of these products has prompted a land-grabbing and land-clearing stampede by foreign investors that has been accompanied by huge road-building networks, processing plants, housing developments, and transport system infrastructure. Writes journalist Nate Spring, "Brazil's agriculture sector grew a sizzling 13 percent in 2017, while the overall economy barely budged. The nation's ability to keep producing new farmland cheaply has given it an edge over rivals and cemented its status as a vital supplier to the world's tables."[59] Today, Brazil's market domination has expanded further due to ongoing US-China trade wars. Brazil has now overtaken the United States and Argentina as the largest exporter of soybeans in the world, with most soy products going to China and the European Union.

Brazil's soybean agribusiness has had devastating impacts on the Amazon rainforest. But with world attention focused on deforestation, the agribusiness industry has moved southward to develop the vast tropical Cerrado savanna region of the country. Compared to the rainforest, the Cerrado savanna has less restrictive environmental protections under the 2012 Forest Code, and the region's development in industrial-scale farming has been actively encouraged by the Brazilian government. Like the Amazonian rainforest, the Cerrado savanna is a unique ecological region in terms of biodiversity and capacities to absorb greenhouse emissions and diminish climate change. As Nate Spring said in 2018, "When farmer Julimar Pansera purchased land in Brazil's interior seven years ago, it was blanketed in tiers of fruit trees, twisted shrubs and the occasional palm standing tall in a thicket of undergrowth." The farmer then mowed down most of the vegetation, set it on fire, and began growing soybean crops. Within a decade the farmer and others in the region deforested an area larger in size than South Korea.[60]

Deforestation in Brazil's Amazon and Cerrado savanna regions is causing massive environmental degradation. Although there was a temporary reduction of rainforest logging from 2005 through 2012, under Bolsonaro's administration deforestation has again rapidly increased. Coming to power with promises of opening rainforest to agribusiness, one of the first things Bolsonaro did as president was to move control over Indigenous-controlled rainforests from the National Indian Foundation agency to the agriculture ministry. He then pulled back funding for forest inspectors, who no longer can monitor illegal logging, and reduced spending on Brazil's national climate-change plan and its commitment to reduce carbon emissions under the Paris Agreement. The Amazon Fund, created and financed by the international community to protect rainforests, was internally restructured and put on hold, further limiting efforts to monitor fire infractions. In August 2019 Bolsonaro organized an event called Fire Day, calling for ranchers to light fires in support of his plans to open public lands to private investors. And then in May 2020 the Supreme Court released a video showing Ricardo Salles, then Brazil's environment minister, calling on the government to further deregulate environmental policy while the country was distracted by its escalating coronavirus pandemic. According to Luiza Lima, spokesperson for Greenpeace Brazil, "Salles believes that people dying in line at hospitals is a good opportunity to move forward on his anti-environmental project."[61]

Under Bolsonaro's extreme-right government, deforestation hit a twelve-year high in 2020 (Figure 6). In response to international condemnation of his tacit approval of burning the rainforest, Bolsonaro claimed it was fake news and that NGOs and environmental groups had lit the fires for attention. Yet despite Bolsonaro's attempts to deflect blame by sending in military troops ostensibly to stop illegal deforestation, the burning continued for months throughout 2019 and 2020. According to a technical report by the Institute for Environmental Research in the Amazon (IPAM), most of the deforestation occurred on undesignated public lands that have been illegally grabbed by ranchers,

FIGURE 6. Destruction by fire of the Amazon rainforest in Brazil.
iStock.

loggers, and miners. As Ann Alencar, director of IPAM, explains,
"When the dry season arrives in the Amazon, these felled trees
will become fuel for burning. This was the main ingredient of
the 2019 fire season."[62] The IPAM report sums up the situation:
"The increase in deforestation in 2020, added to the vegetation
felled in 2019 which was not burned down, leads us to believe
there will be yet another season of intense fire events." The report argues that the escalation of land grabbing of public rainforest requires the federal government to act immediately to stop a
national disaster.[63]

In February 2021 BBC News reported a new twist on the land
grabbing of the Amazonian rainforest. It was discovered that
plots of land—some as big as one thousand football fields—are
being listed for sale in Facebook classified advertisements. These
plots include national parks and protected lands within Indigenous reservations. When approached, Facebook said it did not
have the capacities to police such sales, though it was willing to

work with local authorities in Brazil. However, Ricardo Salles, Brazil's environment minister at the time, noted that the pandemic had made it very difficult for federal agencies to enforce regulations over protected lands, indicating that the government had no intention of taking any action. And this inaction is exactly what buyers are banking on—that once they clear lands for grazing cattle or planting crops, politicians will retroactively legalize their claims on the grounds that open land no longer serves its original purpose of protecting Indigenous sacred homelands and biodiversity. Given a colluding government and Facebook's denial of any responsibility for its participation in illegal property sales, Indigenous groups such as the Uru Eu Wau people face a desperate future. Said Ivaneide Bandeira, head of the environmental NGO Kanindé, "The land invaders feel very empowered to the point that they are not ashamed of going on Facebook to make illegal land deals."[64]

Amazon Rainforests in Global Context

Under President Bolsonaro the rainforest's immediate future, as well as the future of the people and animals that live within it, is in grave jeopardy. However, as the rainforest dies, Brazil's entire agricultural business and its projected long-term profits are in jeopardy as well. Deforestation is causing annual rainfall and water table levels to drop, in turn impacting agribusiness production across Brazil's vast central and southern zones. The impact of felled rainforests doesn't stop at national borders; it also influences weather patterns across the entire South American continent. Globally, the burning of the rainforest is dramatically escalating climate change, as evidenced by satellite data monitoring atmospheric rain and moisture flows. These images show cleared forest areas becoming drier over time. They also show air filled with thick smoke, casting cities in darkness as it climbs high into the sky to drift across continents. Argues Scott Saleska, an expert on the Amazon, "We worry that we will soon cross a threshold of forest loss, a point of no return, after which the water recycling pump will be insufficient to maintain the system and we will see the forest collapse."[65] Adds Carlos Nobre, "Unfortunately, we are already seeing signs of the Amazon turning into a savannah. . . . It's not just theoretical anymore, it's happening already."[66]

The coronavirus pandemic has revealed another global dimension to the burning of the Amazon rainforest. As humans venture deeper into forests to burn and clear, contact with animal life is heightened.[67] This has enabled a global network in trafficked "exotic" animals to wet markets and wildlife markets across Southeast Asia. Rainforest animals caught and sold in a wet market in Wuhan, China, are believed to be the source of the COVID-19 virus. In response to the pandemic, China banned the trade and consumption of wild animals for food and temporarily closed the Huanan Seafood Wholesale Market. But increasing contact between humans and animals is creating a rise in zoonotic diseases, diseases that are passed from animals to humans. According to a UN Environment Programme (UNEP) report in July 2020, zoonotic pandemics are directly attributable to anthropogenic destruction of nature and the global demand to consume meat. "The science is clear that if we keep exploiting wildlife and destroying our ecosystems, then we can expect to see a steady stream of these diseases jumping from animals to humans in the years ahead," said UNEP executive director Inger Andersen.[68]

Extractive capitalism is flourishing in Brazil and across South America on many fronts. Driven by greedy corporations determined to make a profit at any cost, it is also an economic strategy of immense shortsightedness. Deforestation of the rainforest is ultimately going to lead to the drying up of rains and undermine the sustainability of industrial-scale soybean production across Brazil, one of the primary reasons for deforestation in the first place. And deforestation is also contributing to a global market in exotic animals, in turn heightening zoonotic pandemics, which threaten people living in Brazil, the wider continent, and the entire world. Bolsonaro himself contracted COVID-19, but his illness—and the extremely high death figures in Brazil—seem to have made no difference to his policy decisions regarding the Amazonian rainforest, where fires continue to be deliberately lit throughout 2020 and into 2021.

Concluding Comments

The three discussions of Brazil, Australia, and the US underscore the blatant disregard for human and nonhuman life by profit-driven corporations and the politicians who collude with them. This disregard led to the predictable escalation of catastrophic wildfires in recent years. Fires continue to burn and will intensify in scale and destruction as greenhouse gases send temperatures higher, leading to extreme weather patterns of droughts and floods. Many tens of thousands of people, and untold billions of animals, will suffer and die. The violence, callousness, greed, shortsightedness, and deliberate ignorance in denying climate science by those involved in extractive capitalism—be it in Brazil, Australia, or the US—boggles the mind. But perhaps even more astonishing is why ordinary people put up with it. In the next chapter we explore the reasons behind the global rise of antidemocratic leaders and what this rise suggests about the status of global democracy and people's ability to fight against the power of extractive capitalism, denounce antienvironmental policies, and seek environmental justice on behalf of human and nonhuman life.

3

Fire as Weapon

Rising Global Authoritarianism

IT IS NOT COINCIDENTAL that escalating catastrophic wildfires—and climate change in general—are accompanied by a
global rise in antidemocratic governance. I call this trend *free-
market authoritarianism* to explicitly connect neoliberal capitalism with antidemocratic practices such as defunding public education and health, censoring a free press, militarizing the police,
suppressing voting rights, and undermining judicial independence. Around the world, what we are witnessing is an increasing
number of countries, including those claiming to be liberal democracies, tilting toward a concentration of power in a leader or
an elite group who are not accountable or responsible to all their
citizens equally. Writes Ivan Krastev, these political regimes typically have elected leaders who once in power reduce democratic
governance to the "unconstrained power of a majority," such as
in India, Turkey, and Hungary. He adds, "The new authoritarianism fashions itself not as an alternative to democracy but as a
real democracy, one in which a majority governs. A simple Cold
War continuum with democracy on one end and authoritarianism on the other no longer suffices."[1]

Co-opting the rhetoric of democracy and freedom helps veil many authoritarian leaders' deeply antiegalitarian and discriminatory policies and political decision-making. These discriminatory practices take various forms that reflect the unique histories and cultural and ideological complexities of each country. Yet common to almost all of today's free-market authoritarian leaders are two interrelated policy platforms and political strategies: ultranationalism and international isolationism. Ultranationalism promotes the idea of a country's exclusive national identity, often essentialized in racial, ethnic, or religious terms. It is found in political campaign slogans such as "Make America Great Again," "Make Brazil Great Again," and "Make India Great Again." Isolationism promotes the idea of a country's sovereign independence from the rest of the international arena, often mobilizing nativist narratives that evoke racial or civilizational superiority with respect to other nations and peoples. This is very evident in widespread anti-immigrant rhetoric. Isolationism falsely touts a country's ability to thrive without having to cooperate with other countries around global issues such as pandemics, nuclearization, immigration, security, and climate change. We see this in Brexit and the UK's withdrawal from the European Union, and we see it in Trump's withdrawing the US from the Paris Agreement, the Iranian nuclear peace process, and the World Health Organization.

I argue that there is a third common feature in the rise of free-market authoritarian governments that has not yet attracted much attention: antienvironmentalism. As political leaders and big businesses increasingly work together in the pursuit of profits and power, antienvironmental policies (including climate-science skepticism and deregulation of environmental protections) have become essential political tools. Importantly, antienvironmentalism appeals to extreme-right social movements as well as to corporate sectors that together shore up a leader's political support while also ensuring corporate financial backing. This chapter explores the links between free-market authoritarianism, extractive capitalism, and extreme-right political actors, and explores how antienvironmentalism fits into this agenda.[2]

Free-Market Authoritarianism

An antienvironment stance is being taken up by a surge of extreme right parties across Europe, Asia, Latin America, and the Middle East, including far-right governments in Brazil, Australia, and the United States under the Trump administration. As many leaders are discovering, scaling back environmental protections is a winning strategy: it pleases their corporate sponsors while also affirming ultranationalist and isolationist ideologies that are very popular with their extreme-right populist supporters. These leaders form part of the global trend toward free-market authoritarianism. This antidemocratic political shift didn't just arrive out of nowhere; it is deeply connected to late capitalism and the dominance of a hegemonic economic ideology that has been prioritizing profits over people since the 1970s.

Although there are historical parallels, today's authoritarian leaders are notably different from those who emerged in the 1920s and morphed into the fascist and totalitarian Spanish, German, Italian, and Japanese regimes of the twentieth century.[3] What we are experiencing is a new kind of free-market authoritarianism that has its ideological roots deep in conservative thinking and specifically neoliberal values that prioritize economic logic and maximizing profits, normalize patriarchal privilege, and support the imposition of austerity measures that defund such things as public education and public health.[4] Some scholars are calling this new form of oppression "authoritarian neoliberalism," pointing to neoliberal policies that undermine liberal democracy in part by targeting, blaming, criminalizing, and racializing working classes, minorities, and immigrants, particularly since the economic crisis of 2008.[5]

As discussed in Chapter 2, neoliberalism is a radical ideology that pushes a political, social, legal, and economic agenda in which profit-seeking capitalists strive to act without regulatory limitations. It is associated with state policies of deregulation, privatization, dismantling of labor unions, and defunding of public education, health, and welfare services. And it is also connected to re-regulation, the creation of vast transnational legal networks, and a new system of international economic law to serve the interests of global financial capital. In effect, neoliberalism enabled the corporate takeover of regulatory agencies at domestic and international levels to best serve companies' profit-making objectives.[6]

Since the late 1970s, neoliberalism had become the dominant way of organizing economies and societies, though its roots go much further back to the interwar years of the 1930s with the ideas of Austrian economists, most notably Friedrich A. Hayek and Ludwig von Mises. Gradually their thinking began to gain momentum in Europe, eventually influencing economists around the world, including Milton Friedman and Arnold Harberger at the University of Chicago and the overtly racist James M. Buchanan at George Mason University.[7]

Although neoliberal ideology has shifted over time, a constant element is the submission of politics to economic metrics and markets.[8] This requires banks, financiers, and businesses to capture and transform the state to serve their corporate interests. Relatedly, it also necessitates opposing the principles and practices of democracy that seek to limit the power of markets to freely exploit human and nonhuman resources. Notes Wendy Brown, "Throttling democracy was fundamental, not incidental, to the broader neoliberal program."[9]

The result of a decades-long campaign by market fundamentalists to promulgate neoliberal policies has infused the global system with antidemocratic logic, in which the ideas of society and social obligations are denied, the rule of law serves a white economic elite, and authoritarianism is ultimately preferred over representative democracy. This logic crystalized with Chile's dictatorship orchestrated by the United States in 1973, which influenced the thinking of Margaret Thatcher in Britain and Ronald Reagan

FIGURE 7. Hundreds of feet in the air, four climbers from the Rainforest Action Network and the Ruckus Society hang a giant banner off a construction crane on the eve of the mass street protests against the World Trade Organization (WTO), Seattle, 1999. More than 50,000 protesters met to challenge the WTO's exploitative and antidemocratic practices, including labor union members, environmental activists, Indigenous communities, and social justice advocates from the global north and global south. Photo by Dang Ngo / Courtesy of the Rainforest Action Network.

in the US. Since then, neoliberal ideology has slowly been embedded within an international economic system that largely operates behind the scenes and is opaque to most people. This system is financially controlled by the World Bank, the International Monetary Fund (IMF), and other international banks, administered and managed by the World Trade Organization (WTO), and facilitated by a densely woven system of international law, transnational legal networks, global law firms, and public-private trading agreements (Figure 7).[10] And it is furthered through a complex web of tax havens and secretive financial dealings that include off-shore trusts and shell companies that shield massive accumulations of private and corporate wealth from regulatory scrutiny.[11]

Free-market authoritarianism operates on a global scale, evidenced in the parallels between "austerity programmes in the global North, and structural adjustment programs imposed upon the global South by the IMF and World Bank."[12] Structural adjustment programs (SAPs) are economic policies directed at developing countries in the global south and implemented through the IMF and World Bank since the 1980s. They consist of loans to poor countries in exchange for certain conditions that reduce the quality of life for their citizens while making it easier for transnational corporations to move in and develop markets and take advantage of raw materials and cheap labor. SAPs have been severely criticized by Joseph Stiglitz, Nobel laureate in economics, for their predatory and exploitative nature.[13] Both austerity programs in rich counties and SAPs in poor countries intentionally seek to "manage" ordinary people by decreasing funding for public goods while concurrently privileging corporate interests through tax cuts, deregulation, and the watering down of antitrust legislation. Importantly, both economic strategies have created massive socioeconomic inequalities and precarious working classes whose cheap labor can be extracted at minimum costs to corporate investors. These inequalities disproportionately impact people of color and those from lower-income classes.

It's not surprising that we are experiencing a systematic attack on liberal democratic practices, given the shift in power away from centralized state institutions to nonstate economic actors that include dispersed corporate networks, bankers and financiers, and private conservative think tanks. This suggests that in addition to the common critiques of neoliberalism that focus on privatization, deregulation, financialization, and the dominance of market logic, it is important to recognize a deeply rooted ideology within late capitalism that rejects liberal democratic principles of inclusiveness and equality, and at its core is bound up with racial hierarchy and patriarchy.[14] Today there are unprecedented levels of inequality between the masses of working-poor populations and a very few extremely wealthy (typically white and male) individuals.[15] As stated by the critical commentator Stephen Metcalf, "There was, from the beginning, an inevitable relationship between the utopian ideal of the free market and the dystopian present in which we find ourselves; between the market as unique discloser of value and guardian of liberty, and our current descent into post-truth and illiberalism."[16]

Capitalizing on the growing socioeconomic inequalities within and between countries are opportunistic politicians claiming to speak for the downtrodden masses.[17] These masses are often distrustful of mainstream political institutions and are socially and economically insecure. Many are attracted to, and easily manipulated by, far-right social movements as well as charismatic strongman personalities. It follows that the growing cohort of antidemocratic leaders in the US, Europe, Latin America, Africa, and Asia is not an anomaly. Moreover, Trump's actions over his four-year presidency are arguably not as chaotic, impulsive, or idiosyncratic as many would like to think. On the contrary, the rise of Trump, Putin, Salvini, Erdoğan, Duterte, Orbán, and other leaders who inspire cults of personality can be understood as the orchestrated outcome of fundamentalist thinking and neoliberal policies that in many cases have enabled a takeover of state power by global capitalism. Thus, although these leaders may hold different political ideologies and be uniquely charismatic to their supporters, it is the wider practices of collusion between governments and corporations, and their cooperative efforts to dismantle democratic principles and processes, that earmarks the current age as one of global free-market authoritarianism. The formal removal of any one of these leaders—such as Trump—is not as important as how this growing cadre of political elites reinforce and support each other's antidemocratic agendas on a global stage.

In today's global drift toward antidemocracy, individuals running for office rely less on the support of the military and overthrowing governments through violent coups as was the case in the 1960s and 1970s, and more on the creeping insidious power of big business to gain political traction. Of course, some coups still occur, such as the military takeover of Myanmar in early 2021. But today it is more common that extreme-right leaders work closely with the corporate sector, often exploiting their nation's natural resources to buy political backing. This was dramatically illustrated in Brazil with Bolsonaro promising he would open rainforests to agribusiness in his campaign speeches, and in Australia with Morrison promising the coal industry that he would open new, often controversial open-pit mines. In the United States, Trump campaigned loudly on a platform of anti-immigration and national white racism with his coded slogan "Make America Great Again." Much more quietly, he promised businesses he would cut corporate taxes, deregulate environmental protections, and grant industries extraordinary access to deforest, mine, and drill in national parks and public lands. Over his presidency he aggressively and effectively did all these things.

Alarmingly, the global trend toward authoritarianism has been quickened by the COVID-19 pandemic, which started in early 2020. In the name of securing public health, many autocratic leaders declared states of emergency, increased surveillance, cracked down on political and social opposition, weakened the independence of the judiciary, and postponed elections. By mid-2020 this trend was causing international alarm. In an open letter titled "A Call to Defend Democracy," five hundred political and civil leaders, former heads of state, Nobel laureates, and democracy organizations issued a warning that the pandemic was providing a foil for the acceleration of authoritarian governance and dismantling of liberal democratic principles.[18]

Examples of these oppressive governmental actions are everywhere. For instance, in Latin America journalist Anatoly

Kurmanaev argued that leaders "have used the crisis as justification to extend their time in office, weaken oversight of government actions, and silence critics—actions that under different circumstances would be described as authoritarian and antidemocratic but are now being billed as lifesaving measures to curb the spread of the disease."[19] In Europe, Viktor Orbán in Hungary and Jarosław Kaczyński in Poland took advantage of the pandemic to undermine the European Union's commitment to democracy. In Hong Kong, China took bold steps to impose national security legislation to crush prodemocracy opposition, delay elections, and embed control over the territory as the rest of the world was preoccupied with public health emergencies. In the Philippines, Rodrigo Duterte was granted extraordinary military power to fight the pandemic, and then introduced antiterror legislation enabling him to conduct widespread surveillance and imprison many suddenly deemed terrorists in June 2020. And in the United States, as the number of COVID-19 deaths reached a crescendo in July 2020, Trump hinted at delaying the presidential election after sending unmarked federal troops into Portland, Oregon, to quell peaceful protests, two moves that were described as "authoritarian" and "fascist" by commentators from across the political spectrum.[20] Said Michael Abramowitz, president of Freedom House, "We are used to seeing this kind of behavior from authoritarians around the globe, but it is particularly disturbing coming from the president of the United States."[21]

Dying Liberal Democracies

The growing collaboration between ultraconservative politicians and big business is relatively easy to see and understand—at least in hindsight. But what is less obvious is the actual mechanics of how these leaders come to power in the first place. How have a wide range of societies, including those that claim democracy and freedom as signature cultural values, allowed an antidemocratic leader to be their spokesperson?

Disturbingly, many antidemocratic leaders are voted in by democratic electoral processes. Often people don't appreciate the slow dismantling of democratic practices and institutions because there is no single coup or military takeover as was frequently the case in the past. Instead, leaders are voted in and then quietly dismantle the checks and balances associated with liberal democratic governance.[22] As political theorist Boaventura de Sousa Santos notes, the US is as vulnerable to authoritarianism as any other country. The difference, he goes on to say, between antidemocrats such as Trump and dictators is that "the latter start by destroying democracy so they can rise to power, whereas [the former] use democracy in order to be elected, but then refuse to govern democratically and democratically relinquish power. From the point of view of the citizenship, the difference is not that big."[23] The process of transitioning from formal liberal democracy to that of a government with merely a democratic veneer is increasingly evident in many countries around the world. Since the early 2000s, free-market authoritarianism has stealthily worked its way into democratic institutions, shaping and consolidating conservative political perspectives and building an extraordinary global consensus about the benefits of market-driven policy, laws, and governance. This slow chipping away of democratic institutions and values is what the historian Christopher Browning calls the "suffocation of democracy."[24]

Around the world liberal democracies are dying as extreme-right leaders take hold of power. Some of these leaders, often characterized as populist strongmen, openly display authoritarian tendencies that take the form of media censorship, limiting people's rights to protest, defunding public education, militarizing the police, attacking free speech, denouncing public officials, and delegitimatizing the judiciary.[25] Other leaders are less obvious in their antidemocratic practices as they work behind the scenes to corrode and dismantle democratic institutions and regulatory agencies that promote transparency and accountability. Legal scholars such as Javier Corrales refer to this process as "autocratic legalism," whereby a leader is democratically elected to office and then systematically uses, abuses, and ignores laws and constitutional oversight "in service of the executive branch." Through what appear to be legitimate and liberal legal means, autocrats consolidate their political power by extending term limits, undermining public trust in elections, and changing the constitution. According to Kim Lane Scheppele, these leaders "use liberal methods to achieve their illiberal results," coming "to power not with bullets but with law." Notably, she adds, these autocrats often borrow strategies and tactics from each other, creating a worldwide surge of declining liberal democracies.[26]

Sometimes these leaders govern unstable political party coalitions that cobble together various constituencies with overlapping interests coalescing around consolidating power, particularly in the early stages of their term. But eventually all leaders secure their legal control while carefully preserving the veneer of legitimate authority. This is often done by systematically discrediting political competition as well as racializing social opposition and alternative perspectives. Common to many of these governments is the scapegoating of immigrants and ethnic, religious, Indigenous, and LGBTQ communities, who are often blamed for unemployment, crime, and growing inequalities. According to a 2020 report from Freedom House, over the past fourteen years there has been a steady decline in indicators used to measure democratic governance. It explains that "India and the United States are not alone in their drift from the ideals of liberal democracy. They are part of a global phenomenon in which freely elected leaders distance themselves from traditional elites and political norms, claim to speak for a more authentic popular base, and use the ensuing confrontations to justify extreme policies—against minorities and pluralism in particular."[27]

The United States under the Trump administration slid into authoritarianism, whether people recognize it or not. Some scholars have called Trump's leadership "aspirational fascism," while others reject any analogy with classic fascist regimes of the interwar years.[28] Irrespective of the label one uses, this sliding antidemocratic trend has been pushed by the conservative Republican Party for years, well before Trump's presidency. Said Stuart Stevens, a long-standing Republican political consultant, "I spent decades working to elect Republicans . . . and I am here to bear the reluctant witness that Mr. Trump didn't hijack the Republican Party. He is the logical conclusion of what the party became over the past 50 or so years."[29] Many analysts agree that the Republican Party has been intentionally and constitutionally rolling back the norms of representational democracy through such things as "gerrymandering, voter suppression, and uncontrolled campaign spending [that] will continue to result in elections skewed in an unrepresentative and undemocratic direction."[30]

Thus, although Trump's urging on the Capitol insurrection and violence on January 6, 2021, was extremely shocking, it was also not surprising. His false claims of election fraud and explicit attempt to overturn the constitutional electoral process by verbally attacking the chambers of Congress echo a much longer history of conservative thinking that blames immigrants and people of color for claiming their rights and demanding a more inclusive society. Racism, nativism, and xenophobia have now become explicit features of a global extreme-right political pattern. At the same time, some commentators argue that the Capitol riots revealed a new force in US politics. According to Robert Pape and Keven Ruby, specialists in political violence, this new force is "not merely a mix of right-wing organizations, but a broader mass political movement that has violence at its core and draws strength even from places where Trump supporters are in the minority."[31]

Three Common Features of Free-Market Authoritarian Regimes

The fragility of democracy in the United States is evident in other liberal democratic countries around the world that, despite holding elections, are tilting toward an antidemocratic trajectory and promoting an extreme-right agenda. As already mentioned, not all these regimes look like the United States, with each national context reflecting unique histories, values, demographics, political ideologies, religious configurations, and legal and political systems. However, there are common features that all these authoritarian regimes share.

The first common feature is policies of ultranationalism that aggressively promote national sovereignty and independence. Ultranationalism is reflected in campaign slogans such as "Make Poland Great Again" and evidenced in such things as protectionist trade policies, harsh immigration laws, rejection of international human rights norms, and explicit discrimination against minority ethnic, religious, and LGBTQ communities.[32] Significantly, ultranationalism appeals to a core party base who yearn for a return to a mythologized "golden" past or reimagined future steeped in romanticized ideas of cultural and religious homogeneity. Not all forms of ultranationalism work in the same way. In some countries (India, Philippines, China, Brazil, Nicaragua, Turkey, Israel, and so on), ultranationalism often appeals to essentialized cultural identities associated with a specific religion, ethnicity, or political ideology. In other countries (Poland, Hungary, Italy, the US, Britain, Canada, Australia, France, Russia, and so on), ultranationalism typically appeals to the racial superiority of whites over immigrants and people of color. Whatever the basis for galvanizing popular sentiment, ultranationalism is proving to be an extremely potent and dangerous political strategy.

The second common feature among antidemocratic regimes is related to the first: the withdrawal of the country from cooperative multilateralism. Cooperation between states emerged in the post–World War II era, symbolized by the founding of the United Nations in 1945. However, the ideal that countries should work together for the good of the entire world has been steadily undermined since the 1990s as more and more countries act unilaterally in their own best interests. Today we have entered a period of dangerous national isolationism. We see this in a general retreat from global commitments such as denuclearization as powerful countries again enter a nuclear armaments race. We also see it in Trump's withdrawal of the United States from the Paris Agreement on reducing greenhouse-gas emissions in 2017 and from the World Health Organization in 2020. But perhaps most importantly of all, we see it in the extraordinary shift of power away from an international order based on country-to-country diplomacy and cooperation to a neoliberal global order where big business in the technology, pharmaceutical, agribusiness, infrastructure, and energy sectors works directly with governments competing for their services, commodities, and resources. These unilateral commercial arrangements are typically signed behind closed doors and made with little regard for how they may affect people and lands within a country or in adjacent countries or regions.

A third common feature of antidemocratic regimes is antienvironmentalism. As discussed in Chapter 2, denial of climate science by corporations, economists, and conservative politicians has been evident for decades. But as fossil-fuel and mineral reserves become scarcer and food insecurity more evident in the wake of the 2008 economic recession, the pursuit of nonrenewable energy production and agribusiness have become primary concerns for national governments. Land grabbing, deforestation, megadam building, and groundwater pumping have in turn led to depletion of water tables, ocean pollution, leaching of toxins from mining, and destruction of fragile ecological systems. Together these wide-ranging activities reflect a new level of resource extraction since 2010 that is having massively destructive impacts on the planet. Defending the rights of corporations to continue their devastating environmental practices has become one of the main goals of antidemocratic leaders, who depend upon corporate economic support and political clout to remain in office.

In sum, the deepening relationship between free-market authoritarianism and extractive capitalism is also giving rise to the deregulation of environmental laws and protections on a global scale.[33] This has occurred in Brazil and Australia and perhaps most clearly of all in the United States under the Trump administration. Antidemocratic leaders are increasingly emboldened to backpedal on their international commitments to fighting climate change while simultaneously dismantling existing domestic environmental laws forcing corporations to be accountable. This is a dramatically quick reversal of the collaborative spirit surrounding the signing by 197 countries of the Paris Agreement in 2015, pledging to implement domestic environmental protections. The story could not be more bleak: very few countries have put into place new environmental laws since 2015, and many countries have actively dismantled the environmental-protection laws that existed at the time of signing. Even with President Biden coming into office in the United States in early 2021 and declaring a commitment to fighting climate change, his actions are mostly about trying to undo the extensive damage the Trump administration had done. Thus, despite moments of hope, what we are witnessing is a correlation between dying democracies and dying ecological systems that is profoundly disturbing. Together these processes underscore a global surge of unethical behavior that respects neither the rights of other people nor of the life-sustaining ecosystems and landscapes they inhabit.

Deregulating Environmental Protections

Under the Trump administration, the United States took a global leadership role in denying climate science and rolling back environmental protections. But many of Trump's antienvironment policies were put in place under the media radar with little scrutiny by the public. Unlike the 2019 election that ushered in Scott Morrison as prime minister of Australia, the US election in 2016 that brought Trump to power didn't revolve around environmental concerns. Trump focused on race and immigration issues instead. This focus spoke to his Evangelical base in rural and rust-belt areas, though he was voted into office by large numbers of wealthy urban communities as well.[34] Much of Trump's racist messaging was about building a big wall to stop "invading" immigrants on the southern border, and his strategies included attacking women of color, African "shithole" nations, and Muslim "terrorists." Such messaging appealed to a white sector of society that feels disenfranchised and angry about perceived favors granted to people of color.[35]

Trump's stoking white nationalism also included promising to reopen coal mining and other rust-belt industries. In the United States these industries often symbolize an idealized "heartland" identity based on a "colonial and racial imaginary of white settlers and pioneers making a living out of an empty land."[36] This imaginary is infused with values of white masculinity and nostalgia for what is thought of as a golden era, when industries such as mining, logging, and steel and iron plants provided stable livelihoods for working-class men.[37] In these communities, environmental protections limiting extractive industries are often regarded as hostile interference by outside experts and bureaucrats from state and federal agencies. Importantly, Trump's appeal to the racialized insecurities of rural and rust-belt communities strategically dovetailed with his plans to deregulate the energy sector, while also providing constant social media distractions. As commentators note, Trump's "noisy" and chaotic racist rantings stand in stark contrast to his antienvironmental agenda, which was "extremely well-orchestrated. . . . With some exceptions . . . most environmental actions have been unfolding in silence."[38]

It's shocking to see the breadth and depth of environmental-protection deregulation under Trump's four-year presidency from 2016 through 2020. By the end of his term, more than one hundred environmental regulations had been officially reversed, revoked, or rolled back (see the Appendix). These rules relate to

FIGURE 8. Patrick Chappatte, © Chappatte, *The New York Times,*
www.chappatte.com.

air pollution and emissions, drilling and extraction, animals, infrastructure, and toxic substance safety and water pollution. The
rollbacks repealed Obama-era rules on emissions for power plants
and cars, opened Alaskan wildlife refuges and other lands for oil
and gas drilling, and reduced wildlife protections for specific species. According to analysts, "All told, the Trump administration's
environmental rollbacks could significantly increase greenhouse
gas emissions and lead to thousands of extra deaths from poor air
quality each year" (see Figure 8).[39]

The rollbacks represent the reversal of fifty years of environmental protection, according to Dr. Elizabeth Southerland, who
served in the US Environmental Protection Agency (EPA) from
1984 to 2017, before resigning in protest of Trump's deregulation policies. Southerland notes that deregulation "translates

into more pollution of our air, land and water, and allows climate change to go unchecked." She adds that it is economically shortsighted and "comes with a huge price tag: increased health costs; lost workdays as more people get sick; and devastating wildfires, floods and prolonged droughts caused by a warming planet."[40] In tandem with the rolling back of legislation, the Trump administration also removed scientific data from reports issued by the EPA and inserted climate-denial language into reports from the Department of the Interior. This stealth removal of data was furthered by the purging of career scientists employed by the government working to combat climate change. And in his last year of office, Trump used his emergency powers under the COVID-19 pandemic to write an executive order that waived sections of the National Environmental Policy Act. Together these actions added up to a very aggressive attack on the country's ability to adequately address the numerous impacts of global warming.[41]

Similar agendas of rolling back environmental protections have been implemented around the world, many under cover of the pandemic. For instance, in Alberta, Canada, the oil industry requested that the conservative premier, Jason Kenney, reduce required monitoring of toxic spills and pollutants, citing the pandemic and reasons of public health. The Alberta Energy Regulator agreed without any public notice or consultation, and indefinitely suspended approximately twenty requirements for major oil sands producers such as Syncrude, Suncor, Imperial Oil, and CNRL. Said David Spink, a retired government of Alberta employee and former director of air and water approvals, "To my mind it is another blank check to industry and reflects a real lack of priority on/for the environment."[42]

In Brazil and Australia far-right governments also busily implemented antienvironment agendas. One of the first things that Bolsonaro did as Brazil's new president was reduce fines on loggers and farmers for deforestation, arguing that penalties were interfering with economic growth. He simultaneously cut budgets and purged staff in his environmental agency, making it largely ineffectual. In Australia Prime Minister Scott Morrison dismantled initiatives such as the National Energy Guarantee policy, which requires electricity companies to meet emission-reduction targets. Morrison also failed to penalize industrial sites that did not meet greenhouse-gas emissions rules set by the previous government.[43]

Perhaps most blatantly of all, Morrison tried to undermine the 1999 Environment Protection and Biodiversity Conservation Act (EPBC Act), a key piece of Australia's environmental legislation intended to protect and manage significant flora, fauna, ecological communities, and heritage sites. While the EPBC Act was under review, Morrison sought to push through approvals on major developments including mining and deforestation projects before they could be properly assessed for their environmental impacts. Scientists and the public were outraged that this was occurring only months after devastating bushfires had killed or displaced nearly three billion animals, making it "one of the worst wildlife disasters in modern history."[44] When the review of the EPBC Act was published, it confirmed what many scientists, lawyers, and green activists already knew. The EPBC Act was declared "ineffective." The report's opening statement reads: "Australia's natural environment and iconic places are in an overall state of decline and are under increasing threat. The current environmental trajectory is unsustainable." In addition to the report, more than 240 Australian conservation scientists sent a signed letter to Morrison urging that Australia's environment laws be revised to include an effective process of compliance and enforcement. But whether scientists, a concerned public, and the report on the EPBC Act will be able to stop the extinction of many endangered species is seriously in doubt, given Morrison's track record of undermining environmental legislation and pandering to the mining and industrial-farming sectors. As bluntly noted by

Suzanne Milthorpe, a nature campaign manager with the Wilderness Society, "extinction is a choice."[45]

Weaponizing Climate Change

What we are seeing across a range of antidemocratic governments is the weaponizing of antienvironmental and climate policies and practices. Antienvironmentalism joins ultranationalism and international isolationism as political strategies common to free-market authoritarian regimes. Antienvironmental policies signal to the business world (and extractive industries in particular) that a leader is working on their behalf. And they signal to far-right populist supporters an ideological position that evokes ultranationalism, antielitism, and nativism. Putting this differently, antienvironmental policies are fast becoming a political and ideological tool used by extreme-right leaders against their political opposition, be they opposing parties or increasing numbers of ordinary citizens involved in a range of social movements protesting environmental degradation and calling for environmental justice.

The weaponizing of climate change became very evident under the Trump administration. Trump's anti-green agenda was primarily about dismantling environmental protection laws to favor big business. But it also involved belligerently attacking his political opponents over environmental differences. On the domestic front, Trump focused his attacks on politicians in California, a state with a long history of leading the nation in green policies and environmental protections. Trump aggressively argued with Democratic California governor Gavin Newsom over catastrophic wildfires that broke out across the state in 2018, 2019, and again in 2020. As Newsom declared a state of emergency and sought aid from the federal government on numerous occasions, Trump shot back that Newsom had done a "terrible job of forest management" and that he should have swept the forest floor to prevent fires, which was a bizarre, unscientific recommendation. In a tweet on January 9, 2019, the president wrote that billions of dollars had been sent to California and that the fires should never have happened, adding, "Unless they get their act together, which is unlikely, I have ordered FEMA [Federal Emergency Management Agency] to send no more money. It is a disgraceful situation in lives & money!" The former president seems to have conveniently overlooked the fact that 57 percent of California's forests are managed by his own federal agencies. Then in September 2020 Trump visited California at the height of the state's worst wildfire season on record. In a tense meeting with Governor Newsom and California officials, Trump denounced climate science on the warming planet and declared, "It'll start getting cooler. You just watch." He took the opportunity to again blame Newsom for inadequate fire management, knowing that Newsom would have to be polite to receive much-needed federal aid.[46]

Over the course of his administration, Trump took further political swipes at Newsom, such as stopping California from requiring higher car-emissions standards than those at the federal level. Trump also prevented California from opposing federally supported fossil-fuel projects that involve laying down pipelines and infrastructure on California lands. Given California's long history of environmental activism and leadership on progressive environmental laws, these restrictions were deliberately designed to anger Democratic suburbanites in Los Angeles, San Francisco, and San Diego while appealing to his base of Republican supporters in more rural and less cosmopolitan regions. Within California Trump intentionally campaigned across farming communities in the Central Valley by holding a press conference in a Bakersfield airplane hangar in early 2020. To a crowd of thousands, many wearing "Make America Great Again" caps, Trump promised to loosen water regulations that he argued hurt farmers unnecessarily and were based on California's "old science and obsolete studies." These statements greatly appealed to far-right populist ideologies of nativism, anti-elitism, and white nationalism. More profoundly, Trump's dismissal of climate science and the impacts of climate change suggested complete indifference to California's escalating catastrophic wildfires and heightened need for additional federal aid. As temperatures soared, with record highs in August and September 2020, and hundreds of fires besieged the entire state, Trump's threats to defund California's fire efforts dramatically politicized standard federal emergency aid procedures.[47]

On the international front, Trump's antienvironmental agenda was used to appeal to his core supporters at home. By declaring his intention to withdraw the United States from the Paris Agreement in 2017, Trump underscored US national sovereignty and its dismissal of the perils of climate change. This move was a deliberate rejection of a multilateral agreement between 197 state parties (almost every country in the world) to work collaboratively to prevent rising greenhouse-gas emissions. In a Rose Garden speech, Trump stated he would "Make America Great Again" by denouncing world leaders, environmental activists, and those who would try to stop his plans to revitalize coal mining and oil drilling in the American heartland. According to Trump, staying in the agreement could "pose serious obstacles for the United States as we begin the process of unlocking the restrictions on America's abundant energy reserves, which we have started very strongly." Moreover, he added, "foreign leaders in Europe, Asia, and across the world should not have more to say with respect to the US economy than our own citizens and their elected representatives. Thus, our withdrawal from the agreement represents a reassertion of America's sovereignty." And in a final flourish, he ended his speech with this: "It is time to put Youngstown, Ohio, Detroit, Michigan, and Pittsburgh, Pennsylvania—along with many, many other locations within our great country—before Paris, France. It is time to make America great again. (*Applause.*) Thank you. Thank you very much."[48]

Trump's speech picked up the three interlinking prongs to his free-market authoritarian strategy: ultranationalism in his constant refrain to put Americans before global humanity, appealing to his party's racialized sensibility of exceptionalism; isolationism in his withdrawal from international commitments seeking to curb extractive industries, appealing to corporations, bankers, and investors; and antienvironmentalism in his calls to reopen coal mining and reject global commitments to pursue renewable energy, appealing to far-right populists, particularly in the American rust belt.

Trump's interlinking of these political policies—ultranationalism, isolationism, and antienvironmentalism—is an approach that has resonated with other authoritarian leaders around the world. It is evident in the angry response by Bolsonaro to the world's condemnation of his encouraging the burning of the Amazon rainforest that raged for months throughout 2019 and 2020. To help stop Brazil's catastrophic wildfires, President Emmanuel Macron of France offered $20 million in aid on behalf of the G7 (Canada, France, Germany, Italy, Japan, the UK, and the US). Bolsonaro responded by saying Macron had a "colonialist mindset" and should stop interfering in Brazil's internal affairs. After a set of angry exchanges, Bolsonaro added that "respect for the sovereignty of any country is the least that can be expected in a civilized world." This exchange spoke to Bolsonaro's ultranationalist base in its rejection of international help from former European "colonizers" and its defiant affirmation of Brazil's independence from global oversight. Concurrently, it also spoke to Bolsonaro's corporate backers, particularly within the agribusiness and mining sectors, assuring them he would not stop their relentless burning of the Amazonian rainforest. Finally, Bolsonaro's indifference to the environmental consequences of Brazil's catastrophic fires signaled, in effect, a message equivalent to "Go to hell." At this moment, on a global stage, antienvironmentalism was encoded with an ideology of radical extremism praising corporate greed and denouncing collective global futures. Antienvironmentalism, it could be argued, was being aligned with an ideology of antihumanity.

Other antidemocratic leaders share similar antienvironment and anticlimate strategies to those of Trump, Bolsonaro, and Morrison. For instance, Rodrigo Duterte (Philippines), Benjamin Netanyahu (Israel), Narendra Modi (India), and Daniel Ortega (Nicaragua) all have political records linked to climate-science skepticism that conveniently let them promote a wide range of businesses that cause environmental degradation. Among smaller far-right parties that are building across Europe—in Italy, Germany, Finland, France, Poland, and Britain—there is a growing surge of anticlimate sentiment. In Spain, for instance, in the wake of devastating wildfires in 2019, the ultranationalist Vox party objected to any reference to climate change in the government's response to fire victims. Several far-right parties that have a big presence in the European Parliament are also pushing anticlimate agendas and calling for the rolling back of environmental protections. Austria's far-right Freedom Party has openly condemned "climate hysteria" and, along with far-right politicians and journalists in Germany, Britain, and the US, have openly attacked and ridiculed climate activist Greta Thunberg.[49]

Around the world we are seeing growing links between antienvironment sentiments and extreme-right political parties and populist movements. No longer able to completely deny climate-science consensus, political leaders are being forced to take a more aggressive stand against growing political pressures to confront the negative impacts of planetary warming. Under these conditions, antienvironmentalism has morphed from a debate about whether climate change is happening into an ideological position that aligns with ideas of antielitism, nativism, and ultranationalism. Antienvironmentalism is fast becoming "one of the identity-defining features of the far right."[50] In the US, argues economic Nobel Prize winner Paul Krugman, "you can't be a modern Republican in good standing unless you deny the reality of global warming, assert that it has natural causes, or insist that nothing can be done about it without destroying the economy."[51]

Given the rise of free-market authoritarianism, it is reasonable to expect that antienvironment ideologies and policies will intensify in the coming years. We should also expect more overt measures to silence green activists and climate scientists through strategies such as defunding research, media censorship, and smear campaigns, and more drastically through false criminal charges, imprisonment, and even violent attacks. Already in the global south the upswing in environmental conflicts and related deaths of protestors—often called environment defenders—is a major concern and disproportionately involves the suffering of Indigenous peoples and peasant farmers.[52]

Ecofascism

There is another deeply disturbing response to climate change among the few people on the far right who are willing to acknowledge that the planet is warming. This response has deep historical echoes with nineteenth-century conservation movements and their white supremacy roots, and with twentieth-century fascism and ideologies of anti-Semitism, racial purity, and pristine homeland.[53] By blending reactionary ultranationalist beliefs, anti-immigrant rhetoric, and apocalyptic language of imminent ecological collapse, some radical far-right groups are crafting a new ideological convergence. Unlike earlier periods, much of this is occurring behind the scenes through digital communications and social media. As noted by Patrik Hermansson and his colleagues, the rise of far-right social movements and related assertions of violence and hate are transnational phenomena.[54] This ideological convergence is evident in the writings of the shooter who killed fifty-one Muslims in Christchurch, New Zealand, in March 2019 and another shooter who killed twenty-two Mexican immigrants in El Paso, Texas, in August 2019. These men's online manifestos wove together white supremacy and doomsday environmental prophecies that referenced the "extinction" of the "white race" by an overpopulated world.[55] Notably, the first shooter intended his murderous rampage to be a model for others and live-streamed his killings, and the second shooter declared his sympathy for the first in his manifesto.

The prominent climate activist Naomi Klein has said about such events, "I think the only thing scarier than a far right, racist movement that denies the reality of climate change is a far right, racist movement that doesn't deny the reality of climate change, that actually says this is happening." She adds that these groups tend to promote an "abhorrent" ideology that "puts white Christians at the top of the hierarchy, that animalizes and otherizes everyone else, as the justification for allowing those people to die."[56] The journalist Beth Gardiner notes, "It is not hard to see why such ideas are making a comeback. As the relentlessness of environmental calamity—epic fires and floods, escalating extinctions, warming oceans—becomes impossible to ignore, the right needs a way to talk about it. . . . Violent actors are grabbing hold of such ideas."[57]

The weaponizing of climate change and the rising prominence of what some analysts are calling "ecoterrorism" or "ecofascism" is extraordinarily worrying. Climate activist Hilary Moore reminds us that violent radicals don't need to hold formal positions of political power to exert a great deal of influence on societies through social media.[58] How catastrophic wildfires may feature in radical doomsday scenarios, and what political purposes they may be put to by emboldened antidemocratic leaders pushing disinformation and exploiting extreme-right conspiracy theories, is the stuff of nightmares.

Concluding Comments

The global rise of free-market authoritarianism is creating new challenges for those seeking to slow climate change and warming of the planet. These challenges involve confronting the exploitative practices of big business and the extreme-right leaders who support them. But as democratic institutions are slowly chipped away, and in some cases dismantled, it also means that people's ability to push back against corporate power will be reduced as political and legal avenues to effect change are foreclosed. Putting this differently, free-market authoritarian leaders are invested in attacking democratic systems of transparency that call into question the environmental impacts of extractive capitalism on which political leaders economically depend. This conflation of political and economic power, implemented by elites through largely unregulated international finance and legal systems and made sense of by millions of ordinary people through ultranationalist ideology, disinformation, and conspiracy theories circulating through global social media, is historically unique to the twenty-first century. It is occurring in the global south and global north, and as discussed in this chapter, is gaining entrenched political legitimacy, particularly in the context of the COVID-19 pandemic and related looming crises of global unemployment, food insecurity, and poverty.

A feature of the global drift toward antidemocracy is the weaponizing of climate change, and the wildfires it fosters, by leaders to further their political agendas. We see this in the dismantling of domestic laws intended to protect people, animals, and places from the environmental degradation of extractive capitalism. And we see this with countries either withdrawing or disregarding international pledges to lower greenhouse-gas emissions. Antienvironmentalism is fast becoming a signature policy of the radical right, in part because it affirms a country's nationalist independence and in part because it justifies a country's rejection of multicountry commitments to fighting climate change that include making corporations accountable for polluting, poisoning, and killing. Distressingly, ultranationalism and international isolationism are happening at exactly the time that countries need to work together on behalf of the planet's collective humanity to confront climate change and, among other things, reduce wildfires.

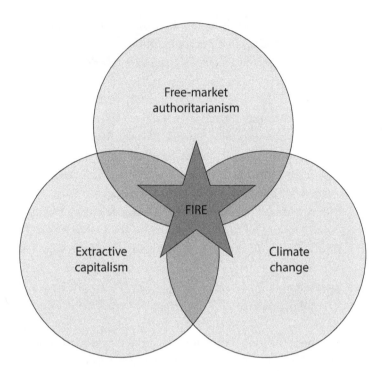

FIGURE 9. Free-market authoritarianism includes a toxic mix of ultranationalism, international isolationism and unilateralism, climate-science denial, and deregulation of environmental protections. © Eve Darian-Smith.

In the previous chapter we discussed the dominance of corporate power and specifically extractive capitalism as the driver of today's global political economy. Linking that conversation to the rise of free-market authoritarianism discussed in this chapter makes explicit the reliance by antidemocratic leaders on big business to finance their political campaigns and keep them in office, as well as on exploiting far-right ideologies of nativism and ultranationalism. Together these two phenomena—extractive capitalism and free-market authoritarianism—contribute to climate change and increasing the number and scale of catastrophic

wildfires. I think of this process as an overlapping and syner-
gistic relationship (Figure 9). It informs today's global political
landscape of antidemocratic regimes and reflects a new phase of
largely unregulated capitalism that is steadily and methodically
consuming the planet and the people, animals, plants, birds, and
fish that inhabit it.

4

Fire as Death

Violent Environmental Racism

AS EXTRACTIVE CORPORATIONS team up with authoritarian leaders—and the impacts of climate change accelerate—we are witnessing a new wave of violent environmental racism in the form of catastrophic wildfires. These wildfires, be they occurring in California, Brazil, or Australia, can be understood as new sites of conflict, suffering, and death that disproportionately impact ethnic and racial minorities. While the Black Lives Matter (BLM) movement has been very significant for highlighting structural racism, an often-overlooked dimension of systemic racial violence is systemic environmental racism, evidenced in the way people of color suffer more than others from the effects of climate change. Be it from polluted rivers, rising oceans, or failed crops, those most immediately and devastatingly affected by environmental degradation are typically people from lower socioeconomic classes, often marginalized on racial, ethnic, and religious grounds.[1] This skewed impact of global warming applies to catastrophic wildfires as well. The disproportionate suffering and death of Indigenous peoples, marginalized peasant farmers, and immigrants linked to out-of-control wildfires is what is explored in this chapter.

I argue that violent environmental racism is a tool and consequence of free-market authoritarianism. Far-right leaders target minority populations—because these communities are often politically weak and can't fight back, and because targeting people of color reinforces a political strategy that speaks to their core supporters and essentialized notions of national identity. In contrast to these specific political strategies, the CEOs, bankers, investors, finance consultants, and corporate lawyers who facilitate extractive industries and their activities don't care who gets hurt, so long as companies turn a profit. Hence, in the converging interests of antidemocratic leaders and corporate businesses, environmental racism can be understood as both an expedient political tool and an unfortunate—though predictable—consequence of the prevailing economic logic.

What makes today's environmental racism different from earlier historical periods is that far-right leaders are now using military force against their own citizens to secure land grabs and defend the toxic practices of extractive industries. As multinational companies seize lands to build dams, lay pipelines, dig mines, and log forests, small local communities, farmers, and Indigenous groups are often driven from their lands and livelihoods by military troops. This is what happened in India in 2020 when the extreme-right pro-Hindu national government introduced three agriculture acts, often referred to as the Farm Bills, that explicitly favor agribusiness. Many small famers felt the new legislation would destroy their livelihoods by making them vulnerable to exploitation by industrial-scale agricultural companies. The Farm Bills incited protests by farmer unions for months, mainly in the states of Punjab and Haryana, which have large Sikh populations. The protests led to a national protest on January 26, 2021, when tens of thousands of farmers marched into Delhi under bombardment of tear gas and water cannon, and met with police violence. According to analysts, the Farm Bills are part of wider political agenda by Prime Minister Narendra Modi and the ruling

Bharatiya Janata Party. "The social and cultural dimensions of the Hindu right's authoritarianism underwrite its unabashedly neoliberal economic agenda. Modi rose to national prominence by implementing the 'Gujarat Model' of politics in his home state, which essentially promotes economic growth by any and all means necessary, including extreme violence."[2]

In the United States, the practice of deploying militarized police against citizens on behalf of corporations is also evident in the numerous protests against the Keystone XL Pipeline and the Dakota Access Pipeline, which would pump enormous quantities of crude oil from Canada into the United States. Beginning in 2011, protests erupted over the laying of pipeline over small farming properties and Native American reservation lands, threatening numerous tribes' clean water and natural resources.[3] The protestors, known as "protectors" of water and sacred lands, fought for years against possible oil spillage and toxic runoffs, and mobilized a global social movement against the projects. Former president Barack Obama halted both pipelines because of the negative impacts on Indigenous sovereign rights and the environment in 2015. But the Trump administration revived the projects, and protests continued at the Standing Rock Sioux Reservation in North Dakota involving hundreds of tribes and green activists. In 2017 Trump deployed the National Guard, who, wearing riot gear and wielding batons, dispersed crowds in what protestors called "a military style operation." Intermittent unrest erupted over months, and a private mercenary firm was brought in to work with the FBI and other government agencies to clear the encampment. After clearing all legal and environmental hurdles, Trump enabled the projects to go ahead. However, President Biden on his first day in office on January 20, 2021, canceled permits to TC Energy for the controversial fourth phase of the Keystone XL Pipeline. This decision was bitterly challenged by twenty-one Republican attorneys general. After months of stalled negotiations, finally, in June 2021, TC Energy announced the termination of the Keystone XL Pipeline. This news was received with much rejoicing by hundreds of environmental justice and Indigenous groups, and by many farmers positioned along the route who had been in legal battles against TC Energy for years. The Native American Rights Fund declared the termination a major victory over fossil-fuel companies, removing "the imminent threat to the tribes, people, and sacred place that stood in the pipeline's proposed path."[4]

The Trump administration's aggressive deployment of armed police against people protesting environmental degradation is unique to the current historical moment, and is extremely alarming. Of course, for decades the US military has seized Indigenous lands and despoiled them with nuclear testing and toxic dumping. As argued by Native American environmentalist Winona LaDuke, the US military's environmental exploitation of Indian country is unparalleled and ongoing.[5] But marking the current moment is the explicit way such exploitation is being carried out by corporations, defended and emboldened by a state military presence. Military power is now actively used by extractive industries whose economic interests have been deemed to be those of the nation as well. And military force is directed not just against Indigenous peoples protesting the taking of their lands but against all citizens who get in the way of extractive industries seeking to log, deforest, plant, mine, drill, extract, unearth, poison, and kill.

This chapter reveals the interconnections between two phenomena that are typically not discussed together. The first is authoritarian leaders' use of militarized police against their own citizens to secure the economic profits of extractive capitalism, often justified by a call for "law and order."[6] The second is accelerating climate change, extreme weather conditions, and the number and intensity of catastrophic wildfires, which disproportionately impact racial and ethnic minorities. I argue that it is vital to see these two phenomena as integrally connected to better understand how environmental racism is implemented and manifested.

The increasing deployment of state military force is key in monitoring these processes emerging in many countries around the world including Australia, the United States, and Brazil. What unites the far-right leaders of these countries is their disregard for democratic principles such as defending their citizens against environmental exploitation, degradation, suffering, and long-term public health impacts. Writes peace activist Vandana Shiva, "Militarization is the shield of corporate globalization, both nationally and globally. At the national level, militarization is becoming the dominant mode of governance. . . . Economic growth is literally flowing through the barrel of a gun."[7]

Australia's Bushfires and
the Suffering of Indigenous Peoples

In 2007 the Australian government implemented the Northern Territory Intervention, also known as the Northern Territory Emergency Response or simply "the Intervention," a military-style operation sent to manage Indigenous communities in remote outback northern Australia. In my mind, perhaps no event more clearly demonstrates the interaction between racialized minorities, profit-driven interests of extractive capitalism, and use of state military force against minorities in support of corporations. Together these activities have contributed significantly to climate change, resulting in catastrophic bushfires that have disproportionately impacted Indigenous Australians.[8]

Racism underscores Australia's history of settler colonialism and continues to inform current state practices governing the nation's Indigenous populations. The colonial legacies of systemic racism are, in the context of climate change and its disproportionate impacts, also environmentally racist. The same sensibility that sees extractive capitalism and its promotion of profits over society's best interests is also tolerant of locking up Indigenous peoples, robbing them of access to their lands, denying them basic human rights, blasting craters into sacred landscapes, and contributing to devastating bushfires that disproportionately impact their communities.

It is impossible to talk about contemporary racism in Australia and how it impacts Indigenous communities without some understanding of the country's colonial history of British settler colonialism.[9] Very briefly, in 1770 the explorer and navigator Captain James Cook evoked the Doctrine of Discovery and laid claim to the island continent. At this time, it is estimated that there existed a population of up to one million Indigenous peoples, constituting more than 250 different nations speaking nearly as many different languages.[10] Despite the clear evidence of Indigenous presence, Cook declared the continent *terra nullius* (empty land) and established the legal basis for a British settlement to be founded. Under Captain Arthur Philip, a penal colony was established in Sydney Cove in 1788. Very swiftly European diseases such as smallpox and influenza wiped out many of the surrounding Aboriginal communities. According to Lieutenant Fowell in 1789, only one year after the colony had begun, "Every boat that went down the harbour found them lying dead on the beaches and in the caverns of the rocks."[11] Those Indigenous peoples that were not wiped out by disease were driven from their lands as the British settlement grew and cleared land for farming. Many were poisoned or hunted and killed if they resisted at all, as most settlers at the time considered Indigenous peoples akin to dingoes, emus, and kangaroos.

Legal, social, and economic forms of discrimination against Indigenous Australians continued well into the twentieth century despite pockets of resistance such as the Australian Aborigines' League, which was started by Indigenous activists in Melbourne in 1934, and the Aborigines Progressive Association, a related Sydney organization that started in 1937. World War II considerably slowed down Indigenous activists' demands for civil and political rights, with the entire country focused first on war participation and then recovery efforts throughout the 1950s. It was not until the 1960s that real change occurred with the establishing of the Federal Council for Aboriginal Advancement, which was heavily influenced by the NAACP and Black civil-rights reforms in the United States. As international media attention condemning Australia's treatment of its Indigenous peoples gained momentum, many Australians enthusiastically passed the 1967 Referendum, which declared that Indigenous peoples must be treated as humans and included in the counting of the Australian population. However, although Australian society grew more culturally diverse and tolerant of racial differences in the 1960s and 1970s, it was still deeply biased against Indigenous peoples and practiced a range of paternalistic and discriminatory laws. One of the most horrifying was the child-removal policy, whereby Indigenous children were forcibly taken from their families and placed in missionary schools and compelled to speak English, dress like white children, and assimilate into white culture. This practice was not entirely abolished until the 1970s. Today those people who suffered such injustices are referred to as the "Stolen Generations."[12]

Indigenous land rights have always been a central issue of conflict between Indigenous and non-Indigenous Australians. One of the main reasons land title is so contentious is that some Indigenous communities hold legal rights to mineral-rich lands that are very attractive to mining companies, and as a result conflict over these lands has been raging for decades. These lands are typically in remote, isolated regions of central and northern Australia that were originally rejected by white settlers as not being viable for farming. In the 1990s Indigenous groups started to slowly gain some political traction and recognition of their land rights. In 1992 the High Court of Australia recognized in *Mabo v Queensland (No. 2)* that Indigenous peoples could hold title over lands and that the common-law doctrine of *terra nullius* did not hold. This decision was considered a landmark at the time and gave rise to the Native Title Act of 1993. However, over the ensuing years, courts have interpreted land claims very narrowly and prevented Indigenous Australians from holding title equal to that held by non-Indigenous Australians. Specifically, the High Court determined in the *Ward* judgment of 2002 that Native title did not give Indigenous peoples rights over resources such as valuable subsurface minerals.[13] This lack of legal equality affirms a sense of white paternalism and superiority over the country's Black citizens, all the while slowly chipping away at what constitutes Indigenous land rights.[14]

Northern Territory Intervention and Grabbing Indigenous Lands

Land disputes between Indigenous Australians and mining corporations frame the Intervention, implemented by the Australian government in 2007. From the perspective of the conservative government at the time, Indigenous peoples were an obstacle to mining companies and needed to be stopped from acting as rights-bearing landholders negotiating in their own best interests and in some cases even blocking corporate takeover efforts.

The Northern Territory is a very large area of Australia with a population of only 220,000, of which about 35 percent are Indigenous. First Australians technically own half the land mass of the territory. Six months prior to the 2007 Intervention, the federal government offered to build homes for some Indigenous groups in exchange for ninety-nine-year leases over tribal lands. Indigenous peoples, including the Yolngu people, rejected the offer, seeing it as a bribe by the government to mine and decimate sacred sites. This prompted the government to take more aggressive action and declare emergency measures to address what they claimed was widespread sexual abuse of Indigenous children, including pedophile rings. Without consulting Indigenous leaders, the government quickly sent the army into the remote Northern Territory. As a result, seventy-three Indigenous towns and encampments were targeted for a range of changes to welfare services, land tenure, and other civil and political rights. In addition to outlawing the use of alcohol, people's welfare checks were partially quarantined, their income supervised, and mandatory medical checks were performed on children to ascertain abuse. Legal actions were exempted from considering customary law and cultural practices, and Indigenous peoples lost their right to manage access permits for non-Indigenous people to enter their local communities.[15]

But the Intervention's most devastating long-term impact was the takeover by the federal government of tribal lands. Indigenous land holdings were initially confiscated under five-year leases, but these turned into leaseholds up to ninety years, to be controlled by the government in return for essential services.[16] The leases were then parceled out to mining companies to dig and remove mountains of soil, extract minerals such as uranium, lower groundwater levels, leach toxins into fragile bushlands, and kill local flora and fauna, all the while contributing to climate change and the conditions in which catastrophic bushfires burn (Figure 10).

The net result of the Intervention was that all sense of dignity and self-determination, including control over traditional lands, was taken away from targeted Indigenous groups. Experiencing the Intervention firsthand, Ali Cobby Eckermann, an acclaimed Australian poet and Yankunytjatjara/Kokatha woman born on Kaurna land in South Australia, writes about the horrifying conditions under which Indigenous communities suffered and continue to suffer. This includes the building of toilet blocks on sacred lands, the enclosing of administrative offices in barbed wire, and the use of doctors without providing for local interpreters. In a poem titled "The Parable," she writes how the "army arrive in their chariots," and "Parents and children race for the sandhills . . . hiding in abandoned cars along the fence line." Eckermann goes on to say that the Intervention "was the moment that any sense of equality and respect, garnered over the previous long

FIGURE 10. Ron Tandberg. © Tandberg. *The Age*, Melbourne Australia 2007.

years by our grandparents and parents, was abandoned by Australia's majority. This was the pivotal moment of division; the moment when the 'freedom of rights' within the 'lucky country' was eroded for Aboriginal peoples, and for other minority groups to come."[17]

Over the years the Intervention has been widely and openly criticized by Indigenous elders and community leaders, civil society organizations and NGOs, the United Nations, and a broad sector of Australia's public, who view it as a failure to meet the real needs of Indigenous Australians. In the many years since its implementation, there is no evidence or reporting of any person being prosecuted for child sexual abuse, which was the original justification for its implementation. In 2012 new policies were introduced to further extend government powers over Indigenous communities under the new Stronger Futures legislation. The new legislation brought in tougher restrictions and introduced the most severe social security penalty in the history of the country—a thirteen-week nonpayment period—for parents whose children do not attend school regularly. The extension incited Malcolm Fraser, the former prime minister of Australia, to declare the legislation one hundred years out of date and "old fashioned white paternalism at its very worst." Despite widespread protests and petitions by Indigenous and non-Indigenous leaders, Concerned Citizens of Australia, and Amnesty International, the Intervention has been extended until 2022.[18]

Slow Violence and Intergenerational Suffering

Although many people acknowledge the failure of the Intervention, there is a limited understanding by white society of the kinds of suffering First Australians experience because of environmental racism. Indigenous peoples live through multiple forms of death under extractive capitalism. They suffer intergenerational health and economic injustices wrought by federal policies and enduring colonial legacies, as is the case with the Intervention. These injustices involve "slow violence" over years and lead to shortened lives, depression, unemployment, suicide, and disproportionately high rates of incarceration.[19] Indigenous peoples suffer again when mining activities undermine their stewardship of lands that they have lived on for approximately 60,000 years. There are numerous instances of mining companies intentionally blowing up sacred sites.[20] And Indigenous peoples suffer yet again because mining contributes to climate change, creating the conditions for catastrophic fires that flare out over thousands of miles to burn across the vast Australian continent. The enormous fires of 2019 and 2020 turned to ash sacred trees and mythologized animal habitats, and smothered with soot thousands of cultural heritage sites containing ancient rock art and cave paintings.

These multiple forms of intergenerational and cultural death are important to acknowledge. Indigenous Australians—like all

FIGURE 11. Protesters are seen during a rally outside the Rio Tinto office in Perth, Australia, Tuesday, June 9, 2020. Rio Tinto detonated explosives in an area of the Juukan Gorge in the Pilbara, destroying two ancient deep-time rock shelters, much to the distress of the Puutu Kunti Kurrama and Pinikura peoples (AAP Image / Richard Wainwright).

human beings—physically die when caught in a fire. But even when they manage to escape bushfires, they can die cultural deaths when the ecological basis of dreamtime stories and ancestral and animal spirits go up in flames. Some Native American writers have called this process "genocide of the mind."[21] The result is that traditional knowledge, stories, languages, memories, and unique relations with creatures and places may be lost forever, including traditional fire-management strategies that have kept their peoples and dreamtime landscapes safe for millennia (Figure 11).[22]

Appreciating the unique suffering and grief of Indigenous . peoples requires rethinking dominant European-based concepts of death. For many of us death happens in a local place such as the hospital or home, marked as occurring at a specific moment

in time. But the multiple physical and cultural deaths Indigenous peoples (and people from other nonwestern cultures) may
suffer transcend local space and linear western time. Moreover,
grief and suffering may include the loss of human and nonhuman
life and a person's past, present, and future relations with various
kinship entities. Understanding the wider frameworks of death
requires us to reexamine our conventional understanding of suffering beyond western concepts when talking about the devastating racialized impacts of climate change.[23] For many millions of
people, environmental racism and suffering span generations, re-
inflicting historical wounds of colonial oppression, reaffirming
marginalization and second-class citizenship that translate into
fewer opportunities to thrive and flourish.

A great deal more can be said about the Intervention—its
horrors, its failings, its robbing Indigenous communities of their
integrity, cultural identity, and land rights as mining companies
move in to dig craters and dump uranium waste. The Intervention cannot be disentangled from long-standing colonial oppression and should be interpreted as part of a continuing history of
land grabbing and exploitation. It is not a coincidence that the
Intervention happened just as Indigenous Australians were gaining modest traction socially, politically, legally, and economically.
This included widespread support by non-Indigenous Australians
for bringing attention to the plight of Indigenous children removed from their homes by the national government up until
the 1970s. At the 2000 Olympic Games held in Sydney, Cathy
Freeman, an Indigenous Australian of Biria and Kuku Yalanji
heritage, was selected to light the Olympic flame in the opening ceremony, which some regarded as a watershed moment in
terms of reconciliation and recognition of past injustices. Freeman then went on to win a gold medal in the 400 meter sprint.
In moving interviews that reached an international audience, she
described how her grandmother had been forcibly taken from
her family. On the legal front, Indigenous Australians were also

making headway that included some legal recognition of their title to land, as well as limited agency to negotiate directly with
companies over royalties for mineral extraction. It is important
to remember that Indigenous Australians' emerging capacities
to fight back against mining corporations is what necessitated—
from the perspective of the conservative government and its corporate allies—the Intervention in the first place.

What is new in this latest episode of colonial oppression is the
federal government's heavy-handed military-style strategies to
"manage" Indigenous peoples on behalf of largely foreign-owned
mining corporations. The deployment of armed police into outback Northern Territory towns can be understood as part of a
wider militarization of the police that has been slowly building
throughout Australia in recent decades.[24] This militarization is
justified on the basis of fighting terrorism and coincides with rising ultranationalist rhetoric that promotes racism, xenophobia,
Islamophobia, and scapegoating of immigrants, many of whom
are held indefinitely in Australia's deplorable immigration detention centers.[25] In a horrifying way, the cordoning off of Indigenous towns with barbed wire and surveillance equipment echoes
the barbed wire surrounding Australia's immigration detention centers. Controlling Brown-skinned "enemies" at the border by locking them up is linked to how the government deals
with the internal issue of landowning Indigenous peoples determined to defend their sacred sites, cultural heritage, and rights
to land stewardship. Both perceived problems are dealt with by
an extreme-right government that promotes divisive racial rhetoric and militaristic strategies. Thinking about the parallels between the Intervention and detention centers reveals deep structures of systemic racism within Australia's governmental policies
and practices. These connections help us to better understand
why many white Australians continue to tolerate environmental racism and explicit land grabbing from the country's original
inhabitants.

Brazil's Deliberate Burning of the Amazon Rainforest

Conflicts similar to Australia's heavy-handed militarized response to Indigenous communities are happening in many places around the world. But nowhere are these battles as intense as in Central and South America, according to data collected by the London-based Business and Human Rights Resource Center.[26] In Nicaragua, Venezuela, Costa Rica, Mexico, Honduras, and Brazil, among other countries, Indigenous groups have clashed with clearing crews hired by agribusiness and mining companies as they ride bulldozers and push excavators deeper into rainforests and woodlands. Approximately eight hundred tribes are fighting with loggers, miners, and ranchers encroaching on their ancestral lands, and the number of violent attacks has risen considerably in recent years.

In Brazil conflict is rapidly escalating between land-grabbing agriculture and mining speculators and Amazonian Indigenous groups and *quilombos* (communities descended from escaped slaves that were constitutionally granted collective ownership to lands in 1988). For many Indigenous peoples this is a reminder of earlier periods of colonial conflict that pushed them off their ancestral lands, as well as the more recent 1970s government development projects that saw mining, logging, and land clearing for cattle grazing and coffee, sugar, maize, and other crops.[27] When President Bolsonaro took office in 2018, he authorized the clearing of rainforests and removal of tribes because he believes they stand in the way of economic development. This was to many people not surprising, since as far back as 1998 he had publicly attacked tribes, stating, "It's a shame that the Brazilian calvary wasn't as efficient as the Americans who exterminated their Indians." And in 2015 Bolsonaro is on record for saying, "The Indians do not speak our language, they do not have money, they do not have culture. They are native peoples. How did they manage to get 13% of the national territory?" His 2018 political campaign was heavily driven by derogatory and racist remarks about

Indigenous peoples, women, and other minorities. The journalist M.B. Sattar writes, "Sadly the Brazilian people were well aware of Bolsonaro's extreme racism when they elected him, just as much as they were aware of his intention to destroy large parts of the Amazon rainforest to make way for capitalist development—whether in the form of mining, logging, or agri-business."[28]

A year after Bolsonaro took office, the removal of rainforest had increased significantly, and burning brush led to widespread fires throughout the region. In response to loud international condemnation for his inaction to stop catastrophic fires in 2019 and again in 2020, Bolsonaro deployed military troops to the rainforest for a month to fight fires and take control of illegal logging and mining activity there. However, according to local reports some troops participated in forcibly moving tribal communities and small farmers from their homelands. This process of dispossession is linked to a renewed plan to colonize and develop the Amazon and to police the northern border of the country with Venezuela. The proposed Rio Branco project sets out to build three new structures: a hydroelectric dam, a bridge across the Amazon, and the extension of a major highway into the northern rainforest. The project builds on fears within Brazil that its national sovereignty is under attack. Many are concerned that the Yanomami tribes on the Brazilian side of the border will unite with those on the Venezuelan side to create an independent Yanomami nation. Bolsonaro and others in the military also fear that Chinese investors in Suriname will try to encroach into Brazilian territory.

Against Bolsonaro's calls for renewed development, environmental groups and Indigenous leaders are calling the project a violation of Indigenous rights that were enshrined in Brazil's 1988 constitution. According to Auricélia Arapiun, a law student and Indigenous leader in Santarém, rather than developing the northern region, the government's plans would "make the Amazon a desert." She concludes, "This is a project of death."[29]

The use of military force to ostensibly fight fires while concurrently clearing lands of small farmers and Indigenous peoples is an ongoing political process with devastating impacts on local communities and ecologically unique rainforest. Even with some troops beating back fires, environmental experts say the deployment was grossly inadequate to stop massive forest destruction. A real response would require restaffing and refunding the environmental protection agencies that Bolsonaro deliberately hollowed out and made ineffective. Adding to this inadequate response, throughout 2020 Brazil was combatting the COVID-19 pandemic, which made accountability of troop activity and law enforcement very difficult. Argues Sueley Araújo, a policy expert who led Brazil's main environmental protection agency before being dismissed by Bolsonaro, "Brazil is becoming an environmental pariah on the global stage, destroying a positive reputation that took decades to build."[30]

One tribe that has been devastated by illegal logging and forest destruction is the Paiter Suruí, who live in a 250,000-hectare territory in the Amazon rainforest in the western part of the country. In desperation, Almir Narayamoga Suruí, chief of the Paiter Suruí Indigenous people, wrote an impassioned letter to Bolsonaro and his government ministers calling for their help. Almir Narayamoga's description of the conditions faced by his community is horrifying: "300 truckloads of timber leave the area daily. The rivers are being contaminated with mercury and cyanide, and the forest dwellers threatened at gunpoint." He goes on: "The Brazilian government has ignored the people's calls for help, and with its silence, is making itself a tacit accomplice to the destruction of the rainforest and the demise of the Paiter Suruí people."[31]

Other Indigenous leaders are also speaking out. Davi Kopenawa, shaman of the Yanomami people, one of the largest Indigenous tribes in Brazil, spoke with journalist Amy Goodman from *Democracy Now*. In the interview Davi Kopenawa said that Bolsonaro doesn't care about protecting the Yanomami people and that his only concern is to make money. Added Kopenawa, "He wants to extract the wealth from the Earth, right from the land where Yanomami people have been living for many, many years." When asked about climate change and both Bolsonaro and Trump calling it a "hoax," Kopenawa said that these leaders were "a sick group" for denying what is clearly taking place. And when asked what message he had for the people of the world, Kopenawa pleaded: "I do not want the president of Brazil to destroy the lungs of our forests, our real Amazon. It is unique. . . . Nowadays, the young people, the youth, is fighting, and they are the ones who will keep on fighting. It is a struggle, a fight so we can keep alive, because without the struggle, we won't live. There will be no forest."[32]

Many Indigenous people have lost their lives in defense of the rainforests. According to a report from Brazil's Indigenous Missionary Council, which seeks to defend Indigenous communities, 135 Indigenous people were murdered in Brazil in 2018.[33] This represents a significant increase in deaths compared to 2017 and not coincidentally is connected in time to Bolsonaro becoming president. On November 1, 2019, a well-known forest defender, Paulo Paulino Guajajara, was ambushed and shot in the head by illegal loggers. His death received national attention and brought enormous sadness to members of the Arariboia Indigenous Territory to which he belonged (Figure 12). Paulo Paulino was a member of a group called Guardians of the Forest, which was set up in 2012 to stop illegal loggers encroaching on tribal lands. And to some degree the forest protectors had been successful. But the nightly patrols are dangerous and increasingly violent, with Paulo Paulino Guajajara's death underscoring just what is at stake for Indigenous peoples. Said Laercio Guajajara, one of the group's coordinators, "I'm proud of the warriors who are continuing the fight, because the Arariboia Indigenous Territory was considered lost. But we're showing the world, the country, that our land isn't lost, that it has an owner."[34]

FIGURE 12. Celia Xakriaba, an Indigenous educator and activist of the Xakriaba people of Brazil, poses with a portrait of Paulo Paulino Guajajara in front of the European Commission headquarters in Brussels, Belgium, November 5, 2019. Leaders of Brazilian Indigenous communities demonstrated outside the Berlaymont building in Brussels, in face paint and headdresses, with a giant flaming tree trunk, asking the EU to stop driving deforestation. (Image by Oliver Hoslet EPA-EFE / Shutterstock.)

In addition to Indigenous communities, rural farmers and workers are also being dispossessed of their lands and livelihoods by agribusiness development. This is the latest phase of dispossession linked to a long history of peasants and small farmers being evicted from their lands in Brazil by elite landlords, who in recent decades have been joined by mining and agriculture industries.[35] For example, 450 landless workers were evicted from the Quilombo Campo Grande camp in the state of Minas Gerais by armed police in August 2020. Despite there being a law in place preventing evictions during the COVID-19 pandemic, police waving rifles and with drones overhead harassed families in a siege that lasted more than fifty hours, eventually forcing the families to flee. The eviction was approved by the state to clear lands for industrial-scale coffee and sugarcane production in a deal with João Faria, one of Brazil's largest coffee producers.

As Brazil's rainforests burn, the European Union and international banks are exerting pressure on Bolsonaro to change his policies and actively intervene to stop the intentional lighting of fires. These efforts may force him to act against loggers, miners, big ranchers, and agribusinesses. But at this point sincere change seems unlikely. Bolsonaro is quite capable, as he has done in the past, of telling the international community what it wants to hear to secure trading agreements and encourage free-market investors. Arguably, this deflection strategy is not unique to Bolsonaro but an example of a wider pattern among free-market authoritarian leaders who want to avert international condemnation of their internal policies. Meanwhile the rainforest continues to burn, people continue to die, and families continue to be evicted by militarized police from lands some tribes have occupied for thousands of years. Forced dispossessions of hundreds of Indigenous groups and thousands of small farmers are becoming more and more common across Brazil and the entire Central and South American region.

California's Wildfires and the Impact on Immigrants

Structural racism underpins and amplifies the differential impacts of climate change. And this is starkly evident in California's enormous industrial-scale agricultural sector in the Central Valley, Salinas Valley, and other farmlands dotted throughout the state. California produces one-third of the nation's vegetables and two-thirds of the nation's fruits and nuts. These are produced by small-range farmers as well as industrial-scale companies that together contribute significantly to climate change in their relentless mining of nonrenewable water resources and polluting with pesticides and fertilizers.[36] The agriculture sector takes an aggressively anti-environmental position. Politically, farmers generally supported Trump when he was president, and their allegiance remains strong in his continuing leadership of the Republican Party. Farmers tend to favor deregulation of environmental protections, but they are also wary of the GOP's tough anti-immigration stance, which undercuts their ability to exploit cheap undocumented workers and can create labor shortages.

Roughly 500,000 to 800,000 farmworkers live in California. The figures are not exact, but most of these farmworkers are Latinx, who make up 92 percent of the workforce; of that number, approximately 77 percent are undocumented.[37] Many come from Mexico, and a large percentage are Indigenous peoples from southern Mexico (Mixtec, Triqui, Maya, and Chatinos). But there are also some farmworkers from Central America who were run off their lands in El Salvador, Guatemala, and Honduras. Many fled gang violence, militias, and corrupt officials. Joining these refugees are growing numbers of Haitian immigrants who fled their country after a devastating earthquake in 2010 and are making perilous journeys up through South and Central America as well. All of these people are searching for better opportunities for themselves and their families.

After surviving the enormous dangers along their journey north, refugees finally end up at the US-Mexico border, where some are deported to their home countries and some held indefinitely in camps while asylum papers are processed in the US courts. Halting immigration into the US was a key plank of the campaign platform for Trump, who pledged to build an enormous wall on the border. Although only eighty miles of new wall were actually built during Trump's four-year term as president, his administration introduced complex laws and protocols, beefed up military police personnel at the border, removed children from their parents as a deterrent, and generally made it extremely difficult for immigrants to gain asylum. However, every year a small number of people do manage to gain unauthorized entry, becoming part of the approximately 2.3 million undocumented Latinx and Indigenous immigrants living in California. Many undocumented immigrants work as gardeners, landscapers, housekeepers, and cleaners, but the majority are employed as farmworkers to pick labor-intensive crops such as grapes, blackberries, avocados, and strawberries.[38]

Undocumented immigrants are extremely careful to avoid interaction with the police and associated threats of deportation. Under the Trump administration, the Immigration and Customs Enforcement (ICE) agency enforced zero-tolerance policies targeting immigrants and those seeking asylum, resulting in a high number of deportations. Jeff Sessions, who served as Trump's attorney general in 2017 and 2018, demanded that local police assist federal agents in arresting and handing over undocumented immigrants to ICE for deportation. California pushed back and introduced an immigrant-sanctuary law that limits local police from having to cooperate with federal immigrant enforcement agents. Much to Trump's anger, the California law was upheld by the Supreme Court in June 2020. In an attempt to further halt deportations, on coming into office in early 2021 President Biden placed a moratorium of one hundred days on such actions. However, his executive order was quickly challenged in court by Drew Tipton, a Trump-appointed Texas judge. Legal challenges will most likely be commonplace as the Biden administration tries to overhaul the draconian and inhumane immigration policies put into place by his predecessor. Such legal uncertainty means that many undocumented immigrants remain extremely wary of any authority figures. For those working in the vast farming fields away from local police, the appearance of gun-pointing ICE federal agents is a constant worry.[39]

Unfortunately, the threat of deportation makes undocumented workers very susceptible to exploitation. Their precarious legal, economic, and social existence makes it extremely difficult for them to unionize or collectively demand better working conditions. But even for those with official visas, being associated with a union or activist group may mean losing your job and only source of income. As one of the most vulnerable working communities, farmworkers toil long days in the fields and often have no access to health and medical care. Pesticides, dust, and heat create appalling labor conditions for farmworkers, whose average life expectancy is only forty-nine years.[40] Farmworkers are exempt from many protections under the Fair Labor Standards Act, including minimum wage guarantees, and are not entitled to overtime pay, sick pay, or mandatory breaks for rests and meals.[41] Some workers sleep in labor camps that resemble prison barracks, others in converted garages and sheds. Whatever the weather conditions, farmworkers go to work to pay their bills. Lucas Zucker, director for the Central Coast Alliance United for a Sustainable Economy, said, "Whether it's wildfire, pandemic, drought, or storm, farmworkers are out in the field." He adds, "There is also often a culture where if you speak up or say you don't want to work, you may be seen as someone who is lazy or doesn't want to work, and you may not be called back for the next harvest."[42]

For both legal and undocumented immigrants, wildfires present a devastating scenario. Legally employers are meant to supply N95 face masks to fieldworkers when air quality is poor, but in practice this rarely happens. Compounding the problem is the lack of face masks because of the COVID-19 pandemic. When fires blazed across California—as they did in August and September 2020—and thousands of homeowners were evacuated because of raining ash, smoke, and extremely poor air quality, farmworkers remained behind to pick strawberries, lettuce, and cherries. According to Juan Reyes, a Spanish-speaking farmworker, "My throat was hurting. There's difficulty breathing. When I got home, my chest was hurting a lot."[43] As temperatures soared above 100 degrees Fahrenheit (38 Celsius) and smoke made it impossible to see, fieldworkers continued their long workdays,

FIGURE 13. Seasonal farmworkers picking and packing strawberries in Salinas, California, 2015. iStock.

risking serious health and respiratory repercussions. In the words of one reporter, "Like the gossamer layer of ash and dust that is settling on the trees in Central California, climate change is adding on to the hazards already faced by some of the country's poorest, most neglected laborers."[44]

If blazing wildfires get too close to the fields and force an evacuation, farmworkers face extraordinary difficulties. People may not own cars and have trouble fleeing disaster zones, finding alternative accommodations, and accessing financial reserves to pay for the immediate needs of food, health, and housing. Sometimes people have trouble understanding emergency information and communicating to emergency services if they do not speak English. Filling out insurance claims, filing new job applications, searching for housing options, new schools, medical care, and so on is very daunting for those with limited financial resources, no medical insurance, and limited English language skills. A recent study shows that the response to the Thomas Fire in California in 2018, which involved 8,500 firefighters and at the time was the second-largest wildfire in California history, completely ignored the needs of undocumented fieldworkers impacted by the blaze. According to the authors of the study, "Society routinely treats undocumented immigrants, as well as other marginalized groups, such as the homeless, as 'less than human, outside the norm, and disposable.'"[45] As a result, emergency relief organizations and policy makers often overlook undocumented workers due to systemic racism and cultural norms about who is deemed a "worthy disaster victim."

Both legal and undocumented immigrants are extremely vulnerable to environmental racism, a form of systemic racism that positions people of color in proximity to such things as lead poison, toxins, pollution, and nuclear waste. In the case of wildfires, the danger lies in smoke inhalation and the potential burning of people's homes, livelihoods, and family members. These dangers are escalating every year as California comes to terms with a new era of year-round wildfires. Against this reality, farmworkers are often disregarded or invisible to those in positions of political power. This is not an accident, but the result of a system structured to silence immigrant's opposition to working in conditions that are extremely hazardous to their health and overall wellbeing. This silencing is vital for extractive industries that need cheap labor to pick crops and keep supply chains flowing, even when the countryside burns.[46] And given the workings of free-market authoritarianism, it's not a coincidence that this silencing was enforced by Trump and continues past his presidential term through the work of his judicial appointees. The end result is the sanctioning of military police to roam borders and fields, threatening immigrants at gunpoint with immediate deportation.

Concluding Comments

The differential impacts of climate change are indisputable. Be it California, Brazil, or Australia, racial and ethnic minorities suffer more and in different ways than the dominant white population. The impacts of catastrophic wildfires, whether de-liberately lit (Brazil) or recklessly allowed to burn (Australia and California), are indicators of deep structural injustices that im-pose the heaviest toll on those least able to bear the burden. Like the COVID-19 pandemic—which climate change has contrib-uted to and exacerbated—the people most likely to suffer from out-of-control fires are disproportionately minorities and people of color. These disproportionate impacts cannot be disentangled from long histories of colonialism that continue to underpin sys-temic racism and racial capitalism occurring on a global scale.

Moreover, minority populations may suffer from climate change in ways not bound by western conventions of time or space. In-tergenerational trauma is often overlooked by Euro-American so-cieties, but for many people the slow violence of environmental degradation over decades cannot be ignored. Gradual leaching of lead poisoning into water systems, accumulations of mercury in fish eaten, smoky air causing deformities in babies born—these are some of the cumulative impacts of human-caused climate change. But intergenerational trauma is not just physical: it involves emo-tional, spiritual, and psychological suffering over decades as well. Loss of memories, languages, stories, and histories embodying special relations to the land adds to the suffering experienced by many Indigenous peoples separated from their ancestral homes. Cultural genocide is not typically something thought about or ac-knowledged in the developed countries of the global north, but for a good portion of the planet's population, emotional and epis-temological violence can be as damaging as physical violence. This means that we need to check our biases and open our minds to the different ways suffering from climate change and catastrophic wildfires may be experienced and borne by others different from ourselves.

The three instances of violent environmental racism explored in this chapter are modest in number and scale compared to the vast environmental injustices resulting from climate change that impact tens of millions of people around the world. Most of these injustices occur in the global south, where the vestiges of colonialism and racism are most evident. Also, these three examples focus on human violence and suffering, leaving a huge gap in the conversation about suffering by animals, insects, birds, and bees that includes the shocking loss of two-thirds of the world's wildlife and biodiversity since the 1970s.[47] However, what the three cases from Australia, Brazil, and California do highlight are some of the ways free-market authoritarianism dovetails with extractive capitalism to produce violent forms of environmental racism.

Specifically, the three examples explored in this chapter show how state militarized force is being mobilized on behalf of extractive corporations to target minority and Indigenous populations. The ensuing confrontations between troops and civilians typically happen on the periphery of the public sphere, beyond media crews, in the remote hinterlands of rainforests, sparsely populated desert landscapes, and vast fields of strawberries and avocados. In Australia, under the Intervention, armed police lock up Indigenous Australians to secure mining companies' access to mineral-rich lands. In Brazil, armed police remove at gunpoint Indigenous and *quilombo* communities from the rainforest for loggers, miners, and ranchers to take over and destroy. And in California, armed federal agents stand ready to deport farmworkers who fail to pick the nation's fruits or protest appalling labor conditions that no human being should be forced to tolerate. The profound tragedy is that many who have left their home countries to flee war and violence in the first place end up being even more vulnerable to environmental racism, exploitation, and premature death. In each case, there are connections between military police deployed to impose law and order at the national border and military police deployed to control Indigenous peoples and racialized minorities within remote interiors. Notably, violence or the threat of violence is always present. These interconnected elements—violence, racism, militarization, suppression, and premature death—are fast becoming the hallmarks of twenty-first-century free-market authoritarianism.

5

Fire as Disruption

Conclusion

AUGUST AND SEPTEMBER OF 2020 were unprecedented in California's history. As many fires raged across California, and night descended over San Francisco in the middle of the day, the sun was shut out by massive smoke clouds hovering over the planet at 40,000 feet. It seemed that the world was coming to an end—an apocalyptic scene experienced by hundreds of thousands of people trying not to breathe in ash, trying to keep warm with the sudden lack of sun, trying to pack up belongings and keep ahead of fast-moving flames. A friend of mine living in the area sent me a text: "Until this morning, I have been taking sunlight for granted." As emergency troops mobilized across the state, planes circled above, dropping pink fire retardant from planes, and thousands huddled in evacuation centers clinging to a few precious belongings, trying to be socially distanced from others as COVID-19 cases surged. For many Californians the wildfire catastrophe was experienced as akin to a war zone.[1]

Fire disrupts normal living. Be it California, Brazil, or Australia, the catastrophic wildfires of 2018, 2019, 2020, and 2021 have changed the landscape for generations, eviscerated fragile ecosystems, and caused the extinction of species that will never return.

They also point to a "new normal," with towns and communities poised to flee as outdoor annual temperatures rise alarmingly. The next round of catastrophic fires is not a matter of if but when. Wildfires are forcing us to rethink and reassess our relationships to other people, animals, plants, trees, and the sun. It is also forcing us to reassess our relationship to governments that have denied climate science and allowed—deliberately or recklessly—such devastation to occur. In the words of University of Sydney professor Chris Dickman, whose study found that three billion animals were killed or displaced in the Australian fires, "I think we've unleashed the demon in climate change. It's very hard to see how we're going to scale things back."[2]

The disruption wrought by wildfires suggests massive changes in how we conduct our everyday lives and how we will have to reimagine our futures. Among many things, it suggests that we will need to think collectively and compassionately about building a new world out of the ashes that includes a different kind of relationship with the planet—one that is not premised on exploiting human and nonhuman resources so that a few individuals can profit. Homo sapiens is one of the species facing extinction if we continue along the same trajectory. To save ourselves, we need to scale back extractive capitalism and the heating of the planet and embrace different priorities, opportunities, and values.

Activists want to disrupt the status quo. That is one of the main points of public protest, to open up conversations and community reflection, and to consider alternatives and what needs to change. That was the purpose of the Black Lives Matter (BLM) movement that erupted around the world in the wake of the killing of George Floyd in Minneapolis in late May 2020, calling attention to structural racism and systemic discrimination against people of color and arguing that when Black lives matter, everyone's life will matter.[3] Disruption was also the purpose of the environmental justice protests that erupted around the world in 2019, calling attention to rising greenhouse gases, environmental racism, and the looming death of the planet. In interesting ways, the BLM movement and the environmental justice movement share similarities in their demands to rethink racial inequalities and discrimination across a wide range of institutions and under increasingly antidemocratic laws and policies, which include heightened militarization of the police in cities and at national borders. Together the two global social movements point to the intimate interconnections among all people and biological life, underscoring that the collective future of humanity rests with eradicating the oppression of marginalized, silenced, and dehumanized individuals.

Demanding Revolution

While catastrophic wildfires raged across Brazil, California, and Australia throughout 2019, environmental activists took to the streets. It was a very busy year of organized protests demanding immediate action by governments on a wide range of environmental justice issues. These protests varied in their attendance numbers, strategies, and explicit political agendas. For instance, in France the predominantly working-class yellow vest movement (*mouvement des gilets jaunes*) teamed up with environmental activists to protest dependence on fossil fuels and exploitation of human and nonhuman resources. In Britain, the global organization Extinction Rebellion took to the streets of Edinburgh, London, and other parts of the country, blocking council chambers, bridges, and busy intersections in a display of mass civil disobedience. Other branches of Extinction Rebellion held protests in New York and Oregon in the United States, Sydney and Melbourne in Australia, and Berlin in Germany. These protests were notable because thousands of citizens who joined the movement came from small communities with no former experience of activism and yet were prepared to be arrested. Extinction Rebellion in the UK was criticized by some who saw it as a white middle-class movement that didn't engage with immediate issues of poverty and inequality faced by some communities of color. These critics also disagreed with its tactics of getting arrested, which assumed the criminal justice system would treat Black and Brown activists the same as white activists. Still, the movement's overriding objective was important in bringing global attention to the estimated one million species of plants and animals facing extinction within decades.[4]

But the most spectacular environmental activism in recent years was the Global Climate Strike, which occurred September 20–27, 2019, coinciding with the United Nations Climate Action Summit and the anniversary of the publishing of Rachel Carson's book *Silent Spring*. This was the third strike of the year by the School Strike for Climate, an international social movement comprised mostly of school students. The movement was founded in 2018 by then 16-year-old Swedish activist Greta Thunberg, whose solo campaign outside the Swedish Parliament in Stockholm demanding the government align itself with the Paris Agreement attracted widespread media attention. Striking every Friday, she inspired school students from around the world to take part in strikes to bring attention to government inaction on addressing climate change. The September 2019 protests took place across 4,5000 locations in 150 countries with an estimated participation of over six million people, including many school students, activists, scientists, and community leaders in what has been called the largest climate strike in world history.[5]

In early 2020 worldwide attention was again focused on climate change during the annual World Economic Forum in Davos, Switzerland. Every year the Davos members-only meeting brings together more than three thousand bankers, financiers, company CEOs, economists, and politicians to discuss and "shape global, regional, and industry agendas."[6] These members represent the economic interests of the world's biggest banks and corporations, and function as the powerhouse of the global political economy. Each member company pays up to 600,000 Swiss francs (approximately US$630,000 in 2020) to be a member, underscoring the importance of participating in the conversations as well as the exclusivity of the event. For decades, members have ignored the crisis of climate change and the massive environmental degradation caused by corporate activities. To appear more receptive to green issues, Greta Thunberg and the UK's Prince Charles were invited as special guests to talk about global warming and carbon dioxide emissions in 2020. Thunberg and Prince Charles's meeting (and formal handshake) symbolized intergenerational solidarity transcending class and status and was welcomed around the world by climate activists.[7]

The teenage activist's message to the global economic elite was explicit and direct, evoking the imagery of fire to underscore the urgency. She accused world leaders of inaction and indifference, saying that the younger generations would not give up without a fight. And she made her passionate speech personal, asking, "I wonder, what will you tell your children was the reason to fail and leave them facing the climate chaos you knowingly brought upon them? That it seemed so bad for the economy that we decided to resign the idea of securing future living conditions without even trying?" In another panel Thunberg called for the "need to bring science into the conversation" along with the voices of young people to build a global conversation about climate change. Prince Charles agreed about the urgency of the situation, and in a keynote address outlined a ten-point plan to guide the global economy to be more sustainable. He emotionally declared, "Global warming, climate change and the devastating loss of biodiversity are the greatest threats humanity has ever faced. . . . What good is all the extra wealth in the world gained from business as usual if you can do nothing with it but watch it burn in catastrophic conditions?"[8]

Thunberg's public appearances at the UN summit in New York in 2019 and at the World Economic Forum in Davos in 2020 attracted worldwide attention. These international events stand in contrast with her other stops during her time in the US to chat informally with schoolkids, community groups, and local environmental justice organizations. These activities reflect her interest in a broader environmental justice movement that is not so much about protecting nature but embracing the deep coexistence of humans with the natural world. Unlike mainstream environmental organizations that want to protect wolves or whales, the environmental justice movement doesn't separate nature from people or think of specific species as discrete entities that need to be saved. Rather, it focuses on holistic interrelations between humans and nonhumans, highlighting the structural inequalities and underlying racism involved in issues such as rising oceans and the forced migration of small fishing communities, chemical poisoning and its impact on children from low-income urban backgrounds, and wildfire smoke and respiratory disease among immigrant farmworkers. This broader picture aligns with more and more people, especially young people, who are increasingly aware of the colonial strategies of land dispossession that ousted poor farmers and Indigenous peoples from their lands to create national parks and conservation spaces devoid of humans.[9]

Thunberg's concern for these broader environmental injustices was evident when she was invited by Tokata Iron Eyes, a young Lakota woman and long-standing critic of uranium mining in Indian Country, to the Pine Ridge Reservation in South Dakota to talk about the climate crisis (Figure 14). A few days later Thunberg also stopped to talk to Indigenous climate activists at Standing Rock, site of controversial climate protests over the Dakota Access Pipeline project. Greta's efforts to raise awareness about climate impacts on Indigenous communities was again evident in her appearance at a UN summit in Madrid in December 2019. She and other youth activists from around the world argued that we need to think differently about climate-change impacts and solutions, to move beyond a focus on carbon metrics

FIGURE 14. Greta Thunberg, age 16 (*left*), sits next to Tokata Iron Eyes, age 16 (*right*), during the panel at the Pine Ridge Reservation in South Dakota, 2019. Photo: Courtesy of Lakota People's Law Project, www.lakotalaw.org.

and economic interests. For instance, they pointed out the structural racism involved when Indigenous communities who are the least involved in producing greenhouse gases are most immediately affected by extreme weather patterns, wildlife extinction, and toxic pollution. And they discussed solutions to climate change that involve science and technology to ensure long-term planetary health, not short-term financial profit. According to Rose Whipple, an Indigenous member of the Santee Dakota in Minnesota and member of the youth team, "The climate crisis is a spiritual crisis for our entire world. Our solutions must weave science and spirituality and traditional ecological knowledge with technology."[10]

Moving conservation beyond the environmental priorities of white Euro-American societies is essential to understanding what is at stake for many millions of people living at the "coalface" or "forest edge" of the climate crisis. For a large sector of the world's population, environmental activism is extremely risky and may even result in death. According to a report by the environmental NGO Global Witness, 212 people were killed in 2019 in their efforts to defend the land and environment, a significant jump from 164 deaths the year before.[11] On average there were more than four murders a week. This made it the deadliest year on record for deaths of land defenders, whose killers are typically not punished. Rachel Cox, a campaigner for Global Witness, said, "Many of the world's worst environmental and human rights abuses are driven by the exploitation of natural resources and corruption in the global political and economic system."[12] The report confirmed that the majority of deaths occurred among activists protesting mining operations, followed by protestors of agribusiness and logging industries. And the largest number of deaths occurred in Colombia, the Philippines, and Brazil, with most of these happening in the Amazon and involving Indigenous peoples (Figure 15). Overall, Indigenous people were disproportionately killed, making up 40 percent of the deaths worldwide. The report also noted that in addition to deaths there were countless cases involving violent attacks, arrests, death threats, and sexual violence against environmental activists, which significantly deter communities from defending their homelands.

These deaths and conflicts over land in the global south are connected to a wide range of environmental conflicts occurring within the global north. Some examples are the Standing Rock "water protectors" demonstrations by Sioux tribes and small farmers over the Dakota Access Pipeline in 2016 and 2017; the Wet'suwet'en First Nation opposition to a Coastal GasLink pipeline project through their territory, which mobilized a wave of civil disobedience blockades against the Canadian government in early 2020; the widespread opposition to Australia's Northern Territory Intervention that effectively imprisons more than seventy Aboriginal communities, denying them the ability to defend their ancestral lands against mining companies until 2022. These and

TOP 5 DEADLIEST COUNTRIES:
Number of land and environmental defenders killed per country in 2019

global witness

1. Colombia	64
2. Philippines	43
3. Brazil	24
4. Mexico	18
5. Honduras	14

Read more globalwitness.org/defenders

FIGURE 15. Image courtesy of Global Witness, www.globalwitness.org.

many more conflicts underscore Indigenous and non-Indigenous groups increasingly joining together to defend fragile hereditary lands, preserve biodiverse ecosystems, prevent catastrophic wild-fires, and slow climate change. For all these marginalized communities, environmental revolution comes at extraordinary costs that should never be discounted or forgotten by those of us living—for the moment—in relatively risk-free circumstances.

Thinking Through Fire

In this book I have argued that we need to think about climate change from the perspective of catastrophic wildfires. This argument takes a cue from environmental justice movements demanding revolution and seeking to disrupt the status quo. Thinking through wildfires as a global calamity forces us to engage with multiple worldviews that don't privilege Euro-American perspectives or profit-maximizing capitalist values. Thinking through fire urges us to move beyond framing environmental crises in ways that speak to bankers, financiers, CEOs, and their authoritarian political partners because they clearly don't care about burning the planet so long as they can make a profit. These economic and political leaders have hijacked conversations about wildfires by denying climate science and the impact of carbon emissions since the 1980s, limiting the terms of engagement to a neoliberal set of economic priorities, and refusing to accept any responsibility for massive environmental suffering and death.

Against the deliberately myopic worldview of capitalists and free-market authoritarian leaders, thinking through fire offers an alternative conceptual framework. First and most important, wildfires make apparent the illusory human/nature divide whereby the environment is somehow understood as disconnected from people. In terms of this divide, some humans—white Euro-Americans—are biologically and morally superior and consequently able to conquer, control, and manage nature. Typically, among the wealthy countries of the global north, great weight is given to new technology as a solution to environmental problems. Thinking that technology can save us suggests that there is no need to change our political, economic, and social systems. It allows to remain in place the logic that underpins extractive capitalism in the form of exploiting cheap labor, and exploiting natural resources through mining, fracking, logging, fishing, and drilling industries. Against this dominant worldview, catastrophic wildfires show us that wealthy elites are not in charge, and that if corporations don't stop emitting greenhouse gases, they won't be able to stop people, birds, animals, and forests from being burned alive any more than they can prevent spirals of ashy smoke rising high into the atmosphere and blocking the sun for days before drifting across continents to drop toxic chemicals onto communities thousands of miles away. The idea that wealthy elites can control nature is simply idiotic. Wildfires show us that every one of us—rich and poor alike—must embrace a relational positionality with the natural world that is bound by respect and interconnection if we want to avoid our own extinction.

Second, thinking through fire makes clear that wildfires, and climate change more generally, do not impact everyone equally. Rising oceans and other environmental crises are having a greater impact on poorer countries in the global south, whose abilities to respond effectively to disaster and risk have been curtailed for decades through a global economic system controlled by the global north through the World Bank and World Trade Organization, and implemented through structural adjustment programs and other neoliberal forms of financial management. These neoliberal strategies build on long histories of colonialism, capitalism, and structural racism and reinforce the disparities of pain and suffering that fires cause, as well as the risk of premature death. It is no coincidence that heightened pain and risk are mostly experienced by ethnic and racial minorities in both the global south and global north. Wildfires also help us to acknowledge that suffering can take different forms and be expressed in ways that may not be familiar to Euro-American societies. This includes the loss of histories, memories, ecological knowledge, and unique relations with the land. Such loss can be intergenerational and experienced over days, years, or decades. Physical, as well as psychological, spiritual, and cultural deaths, are traumatic events for both individuals and communities, and they should not be discounted or devalued when analyzing the short- and long-term impacts of climate change.

Third, thinking through fire points up the need to live as citizens of the world. The rise of ultranationalism and white supremacy, the militarization of national borders and anti-immigrant rhetoric, the withdrawal of nations from international commitments such as the Paris Agreement—all these practices move in the wrong direction, away from recognizing that the future lies in inclusive global cooperation. Catastrophic wildfires are not constrained by national borders, nor should be our thinking about how to find solutions to stop them. This suggests that we must take seriously multilateral international commitments to reduce greenhouse gases that serve a global humanity, as well as global grassroots movements that can facilitate cross-cultural conversations and cooperative community-to-community frameworks.

Last, catastrophic wildfires that have been facilitated by authoritarian leaders connect the political erosion of liberal democratic principles with the corrosion of the environment. As far-right leaders roll back environmental protections and promote extractive industries such as logging, mining, and industrial-scale agriculture, they are also rolling back elements of democracy such as a free press, nonpartisan judges, public health programs, public education, and the right to vote and protest injustice and suppression. The call for law and order and the mobilization of militarized police are common among authoritarian leaders anxious to prevent public demonstrations against them. Violently suppressing citizens opposing the government is deeply entangled with leaders defending extractive industries through state-backed military and police force. Antienvironmentalism has become a marker of extreme-right parties and political agendas as well as a common theme among the radical media and movements that support them. This is the distinct and terrifying profile of free-market authoritarian regimes that are on the rise globally.

For people who have not yet experienced the kinds of catastrophic wildfires that have burned our planet in recent years, it may be hard to think through fire. But everyone has in some way been impacted by the COVID-19 pandemic, so it's easier to start with that shared experience. The parallels between catastrophic wildfires and the pandemic are real and visceral. Both emerge out of the same economic and political global conditions of gross inequality that have been shaping the world for the past fifty years. As the COVID-19 pandemic unfolds and its disproportionate impacts materialize over the coming years, we must take it as a wake-up call for what looms on the horizon in the form of an escalation of fires as the planet warms and oceans rise.

For those of us privileged enough to have been able to shelter in place, the pandemic should help us appreciate that we do not have to travel so much, shop so much, waste so much, ignore others so much, or be so self-absorbed. Looking up from the disruption of our daily lives, we may have seen the massive inequalities and racial injustices so prevalent in our societies, where for many millions of people there was no home to shelter in, no access to health services, and no government aid. Or, instead, with the accessibility of a vaccine, we may return to life before the pandemic. We can continue to deny scientific evidence, abdicate responsibility for our futures, allow greedy corporations to burn forests and pollute the air we breathe, and tolerate authoritarian leaders and their antidemocratic policies that are quickly taking away our collective ability to act. Whatever we do, there is no avoiding that our planet is literally and metaphorically on fire.

ACKNOWLEDGMENTS

This book was written while in pandemic lockdown but reflects my thinking on numerous issues that I have been engaged with for decades. Writing it provided me an outlet for my frustration and despair as death tolls mounted from COVID-19 and dangerous political tweets filled social media. Waking up early in the morning cool and working on this book helped anchor days that otherwise rolled by in monotonous repetition, interrupted by intense weeks filled with sirens and ash raining from orange skies. I am deeply appreciative of my status and economic advantages that allowed me to think and write under such extraordinary circumstances.

With the writing of any book there are many thanks to give. The support of mentors and colleagues over decades washes over this book, especially the friendship and inspiration of Peter Fitzpatrick, Sally Falk Moore, Wes Pue, Sally Engle Merry, Jane Collier, and George Collier. A very special thanks to my mentors back home in Melbourne, Donna Merwick and Greg Dening, for taking me under their wing when I was an undergraduate student, introducing me to Foucault, and encouraging me to think big. While many of these senior scholars are no longer alive, their compassion and generosity continue to guide my thoughts.

Warm thanks to my colleagues in the Department of Global and International Studies at the University of California, Irvine, who helped keep me sane. Special thanks to Shauna Gillooly and Lucy Garbett—two wonderful junior colleagues who read and gave valuable feedback on the entire manuscript. Thanks to Michelle Lipinsky, Naja Pulliam Collins, and especially to Stacy Eisenstark, who were so very supportive of this project from the start, and to my wonderful editor Marcella Maxfield and assistant Sunna Juhn at Stanford University Press. Thanks also to the thoughtful reviewers of the manuscript. Huge thanks to the medical personnel, ambulance drivers, firefighters, caretakers, and all essential workers for helping everyone breathe as COVID-19 lockdowns and smoke from the worst wildfires in California's history kept the privileged—including me—safely indoors. Finally, my most heartfelt thanks go to my family. Thanks to mum and dad, always. Thanks to my loving and inspirational partner Philip McCarty and our beautiful, crazy teenagers Ellie and Sam for putting up with me. Last, but not least, thanks to my twin sister Corinna, who really is my better half.

APPENDIX

A brief list of some of the most destructive of the rollbacks under the Trump administration (2016–20) gives a sense of what has been happening with little public awareness or knowledge. For a full list, see the *New York Times* website.[1]

- Loosened a Clinton-era rule designed to limit toxic emissions from major industrial polluters
- Revised a program designed to safeguard communities from increases in pollution from new power plants to make it easier for facilities to avoid emissions regulations
- Weakened an Obama-era rule meant to reduce air pollution in national parks and wilderness areas
- Revoked an Obama executive order that set a goal of cutting the federal government's greenhouse-gas emissions by 40 percent over ten years
- Repealed a requirement that state and regional authorities track tailpipe emissions from vehicles on federal highways
- Lifted the ban on drilling in the Arctic National Wildlife Refuge
- Rescinded water pollution regulations for fracking on federal and Indian lands

- Scrapped a proposed rule that required mines to prove they could pay to clean up future pollution
- Withdrew a requirement that Gulf of Mexico oil rig owners prove they can cover the costs of removing rigs once they stop producing
- Loosened offshore-drilling safety regulations implemented by Obama after the 2010 Deepwater Horizon explosion and oil spill in the Gulf of Mexico
- Lifted an Obama-era freeze on new coal leases on public lands
- Weakened the National Environmental Policy Act, one of the country's most significant environmental laws, in order to expedite the approval of public infrastructure projects such as roads, pipelines, and telecommunications networks
- Relaxed the environmental review process for federal infrastructure projects
- Revoked an Obama executive order promoting climate resilience in the northern Bering Sea region of Alaska, which withdrew local waters from oil and gas leasing and established a tribal advisory council to consult on local environmental issues
- Changed the way the Endangered Species Act is applied, making it more difficult to protect wildlife from long-term threats posed by climate change
- Reversed an Obama-era rule that barred using bait, such as grease-soaked doughnuts, to lure and kill grizzly bears, among other sport-hunting practices that many people consider extreme, on some public lands in Alaska
- Rolled back a roughly forty-year-old interpretation of a policy aimed at protecting migratory birds, potentially running afoul of treaties with Canada and Mexico
- Scaled back pollution protections for certain tributaries and wetlands that were regulated under the Clean Water Act by the Obama administration

- Revoked a rule that prevented coal companies from dumping mining debris into local streams
- Withdrew a proposed rule requiring groundwater protections for certain uranium mines. In 2020 the administration's Nuclear Fuel Working Group proposed opening 1,500 acres outside the Grand Canyon to nuclear production

NOTES

Preface

1. "Secretary-General Calls Latest IPCC Climate Report 'Code Red for Humanity,' Stressing 'Irrefutable' Evidence of Human Influence," United Nations, https://www.un.org/press/en/2021/sgsm20847.doc.htm.

2. Fionnuala Ní Aoláin, the United Nations Special Rapporteur on the promotion and protection of human rights and fundamental freedoms while countering terrorism, cited in *New York Times*, March 30, 2020.

3. Raza Saeed, "COVID-19 and the Continuity of the Familiar," *Critical Legal Thinking*, March 21, 2020, https://criticallegalthinking.com/2020/03/21/covid-19-and-the-continuity-of-the-familiar/.

4. Osagyefo Amoatia Ofori Panin Okyenhene, "The Consequential Effects of COVID-19 on the Climate Crisis," *IPS News*, June 1, 2020, http://www.ipsnews.net/2020/06/consequential-effects-covid-19-climate-crisis/.

Chapter 1: Fire as Omen

1. Elizabeth Kolbert, *The Sixth Extinction: An Unnatural History* (New York: Picador, 2015).

2. Greta Thunberg, *Our House Is on Fire: Scenes of a Family and a Planet in Crisis* (New York: Penguin Books, 2020). See also Naomi Klein, *On Fire: The (Burning) Case for a Green New Deal* (New York: Allen Lane, 2019).

3. Daniel R. Wildcat, *Red Alert! Saving the Planet with Indigenous Knowledge* (Golden, CO: Fulcrum Publishing, 2009, 109–111).

4. Candace Sutton, "Inside a Raging Beast: What It's Like Inside a Bushfire," *NewsComAu,* October 23, 2013, https://www.news.com.au /national/inside-a-raging-beast-what-its-like-inside-a-bushfire/news -story/ddf8111b9ec1fe6212d23daa0e4d2621.

5. However, researchers are now beginning to understand the long-term impacts of fires that can harm humans and ecological systems in a myriad of ways, such as contaminating urban water supplies for over a decade (see Henry Fountain, "Wildfires Threaten Urban Water Supplies Long After the Flames Are Out," *New York Times,* June 24, 2021). Other long-term impacts materialize as respiratory and cardiovascular health complications, particularly in the elderly. One of the problems in assessing fire-based health outcomes is that smoke may drift over communities far from the original fire and materialize as respiratory problems months, if not years, later. See Bob Henson, "Silent Calamity: The Health Impacts of Wildfire Smoke," *Yale Climate Connections,* May 21, 2021, https://yaleclimateconnections.org/2021/05/silent-calamity-the -health-impacts-of-wildfire-smoke/.

6. Rob Nixon, *Slow Violence and the Environmentalism of the Poor* (Cambridge, MA: Harvard University Press, 2011).

7. Extensive evidence shows that the COVID-19 virus was very selective in terms of class, ethnicity, race, and age, with data demonstrating that it impacted poorer communities of color overwhelmingly. Wealthy (predominantly white) communities were better able to "shelter" from the COVID-19 virus and work from home, and were not forced to go out to do frontline service jobs such as cleaning, caring for others, or working at supermarkets, restaurants, and so on. The disproportionate impact of the pandemic reinforces ideologies of white superiority and the notion that some people are less valuable and more disposable than others. See Eve Darian-Smith, "Dying for the Economy: Disposable People and Economies of Death in the Global North," *State Crime* 10, no. 1 (2021): 61–79.

8. Ruth Wilson Gilmore, "Race and Globalization," in *Geographies of Global Change: Remapping the World,* ed. R. L. Johnstone, Peter J. Taylor, and Michael J. Watts, 261–77 (2nd ed.; Oxford, UK: Blackwell, 2002).

9. Racial capitalism refers to a process in which white individuals and institutions use nonwhite people to acquire social and economic value. The term was first coined by Cedric Robinson, who argued that racism was already apparent in feudal times and formed the basis for modern capitalism and its systems of racialized oppression and exploitation that endure into the contemporary era. See Robinson, *Black Marxism: The Making of the Black Radical Tradition* (London: Zed Press, 1983).

10. Hence the fire referred to by scholars of race in the context of the civil rights movement and contemporary police violence is historically, politically, and metaphorically connected to the wildfires that are the focus of this book. See James Baldwin, *The Fire Next Time* (New York: Vintage, 1992); Elizabeth Hinton, *America on Fire: The Untold History of Police Violence and Black Rebellion Since the 1960s* (New York: Liveright, 2021); and Don Lemon, *This Is the Fire: What I Say to My Friends About Racism* (New York: Little, Brown, 2021).

11. Laura Pulido argues that often overlooked in scholarship on environmental racism is a "spatiality of racism." To better understand how and why different demographic groups occupy different places and neighborhoods, she points to multiple forms of racism working together. Her focus on white privilege reveals how white communities engage in systemic racism through white flight and zoning laws that may be practiced unconsciously and don't necessarily involve explicit hostility to nonwhites. See Laura Pulido, "Rethinking Environmental Racism: White Privilege and Urban Development in Southern California," *Annals of the Association of American Geographers* 90, no. 1 (2000): 12–40; Dorceta Taylor, *Toxic Communities: Environmental Racism, Industrial Pollution, and Residential Mobility* (New York: New York University Press, 2014); Ingrid R.G. Waldron, *There's Something in the Water: Environmental Racism in Indigenous and Black Communities* (Halifax, NS: Fernwood Publishing, 2018); and Michael Méndez, Genevieve Flores-Haro, and Lucas Zucker, "The (In)Visible Victims of Disaster: Understanding the Vulnerability of Undocumented Latino/a and Indigenous Immigrants," *Geoforum* 116 (2020): 50–62, 51.

12. Berch Berberoglu, *The Global Rise of Authoritarianism in the 21st Century: Crisis of Neoliberal Globalization and the Nationalist Response* (London: Routledge, 2020).

13. Corey J.A. Bradshaw and Paul R. Ehrlich examine the similarities of climate politics and what Australia and the US can each learn from the other in their book *Killing the Koala and Poisoning the Prairie: Australia, America, and the Environment* (Chicago: University of Chicago Press, 2015).

14. International Monetary Fund World Economic Outlook, 2020, https://www.imf.org/external/datamapper/NGDPD@WEO/OEMDC /ADVEC/WEOWORLD.

15. In 2017 the GDP or per capita income in the United States was approximately $60,000, Australia $50,000, and Brazil $15,000, https:// www.worldometers.info/gdp/gdp-per-capita/.

16. Umair Irfan, "The US, Japan, and Australia Let the Whole World Down at the UN Climate Talks," Vox, December 18, 2019, https://www.vox.com/energy-and-environment/2019/12/18/21024283 /climate-change-cop25-us-brazil-australia-japan.

17. In May 2021 the Biden administration defended in federal court the Willow project, a Trump initiative allowing for massive drilling on Alaska's North Slope that is projected to produce over 100,000 barrels of oil a day for the next thirty years. Not surprisingly, this project has been bitterly opposed by climate activists. The Biden administration also backed Trump's earlier decision to grant oil and gas drilling leases on federal land in Wyoming, and to allow crude oil to continue to flow through the Dakota Access Pipeline, which doesn't yet have a federal permit. Despite his climate pledge, Biden faces a morass of political, legal, and technological difficulties that make it hard to put into action his climate agenda, which includes cutting fossil-fuel pollution and mitigating the impacts of climate warming. See Liz Ruskin, "Biden Administration Throws Support Behind Massive Willow Oil Project," Alaska Public Media, May 28, 2021, https://www.alaskapublic.org /2021/05/27/biden-administration-throws-support-behind-massive -willow-oil-project-in-npr-a/; Nichola Groom, "Biden Administration to Resume Drilling Auctions in Setback to Climate Agenda," Reuters, August 31, 2021, https://www.reuters.com/business/energy/biden-ad ministration-takes-steps-resume-oil-gas-drilling-auctions-2021-08-31/; and "The Biden Plan for a Clean Energy Revolution and Environmental Justice," https://joebiden.com/climate-plan/.

18. Paul Feiner, letter to the editor, *New York Times*, May 6, 2021, https://www.nytimes.com/2021/05/06/opinion/letters/liz-cheney-republicans.html; Jonathan Chait, "Liz Cheney Wants to Save American Democracy from Trump's Next Coup," *New York Magazine*, May 5, 2021, https://nymag.com/intelligencer/2021/05/liz-cheney-save-democracy-republican-trump-coup-insurrection-purge.html.

19. *Economist*, "In America, Republican-Led States Are Rolling Back Electoral and Civil Liberties," April 19, 2021, https://www.polisci.washington.edu/sites/polisci/files/documents/news/in_america_republican-led_states_are_rolling_back_electoral_and_civil_liberties_the_economist.pdf; Tom Davis cited in Trip Gabriel, "Virginia GOP's Choices for Governor: 'Trumpy, Trumpier, and Trumpiest,'" *New York Times*, May 7, 2021, https://www.baltimoresun.com/news/nation-world/ct-aud-nw-nyt-virginia-governor-gop-choices-20210507-afi5xd6ijnfklj2d7idswkmlqe-story.html.

20. For instance, Governor Greg Abbott of Texas deliberately blamed—incorrectly—electricity power failures due to freezing weather conditions on cost-effective renewable wind power and green activists in February 2021. See Katie Shepherd, "Rick Perry Says Texans Would Accept Even Longer Power Outages 'to Keep the Federal Government Out of Their Business,'" *Washington Post*, February 19, 2021, https://www.washingtonpost.com/nation/2021/02/17/texas-abbott-wind-turbines-outages/.

21. One of the more bizarre stories pushed by US Representative Marjorie Taylor Greene (R-GA), a staunch Trump loyalist openly praised by Trump, was that California's 2018 Camp Fire, which killed eighty-four people, was caused by laser beams from space in a program she claimed was started by Jewish people. For this and other outrageous right-wing conspiracy theories, she was removed from sitting on any committees in the House of Representatives in February 2021. However, most Republican congressional members supported her right to tell such explicit racist lies and refused to denounce her or hold her accountable in any way. See Josh K. Eliott, "'Jewish Space Lazer' Among Wild Hoaxes Backed by GOP's Marjorie Taylor Greene," *Global News*, January 29, 2021, https://globalnews.ca/news/7607501/marjorie-taylor-greene-jewish-space-laser/; BBC, "Marjorie Taylor Greene: US House

Votes to Strip Republican of Key Posts," February 5, 2021, https://www
.bbc.com/news/world-us-canada-55940542. See also Mike Rothschild,
*The Storm Is upon Us: How QAnon Became a Movement, Cult, and Con-
spiracy Theory of Everything* (New York: Melville House, 2021).

22. Cited in Anna Jean Kaiser, "Is the Amazon Really 'the Lungs'
of Planet Earth? No, It's More Like Our Sink," *Chicago Tribune*, August 27, 2019, https://www.chicagotribune.com/nation-world/ct-nw
-cb-amazon-fires-explainer-20190827-oneqsy6c6nfuxb33bictnewnbm
-story.html.

23. The "biotic pump" theory argues that natural forests act as a
pump that draws moisture inland and sets up its own wind and weather
patterns. See Louis Verchot, "The Science Is Clear: Forest Loss Behind
Brazil's Drought," *Forest News*, January 29, 2015, https://forestsnews
.cifor.org/26559/the-science-is-clear-forest-loss-behind-brazils-drought
?fnl=.

24. Cited in Drew Kann, "Extreme Drought and Deforestation
Are Priming the Amazon Rainforest for a Terrible Fire Season," CNN,
June 22, 2021, https://edition.cnn.com/2021/06/22/weather/brazil
-drought-amazon-rainforest-fires/index.html.

25. "Brazil's Amazon: Deforestation 'Surges to 12-Year High,'"
BBC News, November 30, 2020, https://www.bbc.com/news/world
-latin-america-55130304.

26. "The Destruction of the Amazon, Explained," *The Week*, September 1, 2019, https://theweek.com/articles/861886/destruction-ama
zon-explained.

27. Alistair Gee and Dani Anguiano, *Fire in Paradise: An American Tragedy* (New York: W.W. Norton, 2020).

28. This essay appears as chapter 3 in Mike Davis, *Ecology of Fear:
Los Angeles and the Imagination of Disaster* (New York: Vintage, 1999),
93–147.

29. Davis, *Ecology of Fear*, accessed February 1, 2021, https://long
reads.com/2018/12/04/the-case-for-letting-malibu-burn/.

30. David Wallace-Wells, *The Uninhabitable Earth: Life After
Warming* (New York: Tim Duggan Books, 2019), 74.

31. Quoted in Ivan Penn, "PG&E Ordered to Pay $3.5 Million Fine
for Causing Deadly Fire", *New York Times*, June 18, 2020, https://www

.nytimes.com/2020/06/18/business/energy-environment/pge-camp
-fire-sentenced.html.

32. Carolyn Kormann, "When Will Australia's Prime Minister Accept the Reality of the Climate Crisis?," *New Yorker*, January 15, 2020, https://www.newyorker.com/news/news-desk/when-will-australias
-prime-minister-accept-the-reality-of-the-climate-crisis; and Damien Cave, "The End of Australia As We Know It," *New York Times*, February 15, 2020, https://www.nytimes.com/2020/02/15/world/australia
/fires-climate-change.html.

33. McLauchlan, Kendra K., et al., "Fire as a Fundamental Ecological Process: Research Advances and Frontiers," *Journal of Ecology* 108 (2020): 2047–69.

34. Jason Beaubien, "With Their Land in Flames, Aboriginals Warn Fires Show Deep Problems in Australia," NPR News, January 11, 2020, https://www.npr.org/2020/01/11/795224932/with-their-land-in
-flames-aboriginals-warn-fires-show-deep-problems-in-australia.

35. Cited in Leah Asmelash, "Australia's Indigenous People Have a Solution for the Country's Bushfires," CNN, January 12, 2020, https://
www.cnn.com/2020/01/12/world/aboriginal-australia-fire-trnd/index
.html.

36. Cool-fire burning is a controlled form of land management. Small fires are lit, typically by hand, to burn the debris of leaves on the ground, all the while retaining moisture, nutrients, and seeds in the soil for quick rejuvenation. See Courtney Fowler, "Kakadu National Park: Traditional Burning Methods and Modern Science Form a Fiery Partnership," ABC News, August 12, 2016, https://www.abc.net.au
/news/rural/2016-08-12/traditional-owners-fire-management-kakadu
/7730254.

37. Alexis Wright, "Want to Stop Australia's Fires? Listen to Aboriginal People," *New York Times*, January 15, 2020, https://www.nytimes
.com/2020/01/15/opinion/australia-fires-aboriginal-people.html.

38. Nylah Burton, "Invasion Day Is a Day of Mourning for Indigenous Australians. The Bushfires Make This Year Extra Poignant," Vox, January 24, 2020, https://www.vox.com/identities/2020/1/24
/21080027/invasion-day-australia-aboriginal-indigenous-torres-strait
-islander.

39. Livia Gershon, "The Global Suppression of Indigenous Fire Management," JSTOR Daily, October 12, 2020, https://daily.jstor.org/the-global-suppression-of-indigenous-fire-management/.

40. Stephen J. Pyne, *The Pyrocene: How We Created an Age of Fire, and What Happens Next* (Oakland: University of California Press, 2021).

41. Cited in Carla Marinucci, "Brown: California Fires Show 'the Horror' World Will Face from Climate Change," Politico, October 28, 2019, https://www.politico.com/states/california/story/2019/10/28/brown-california-fires-show-the-horror-world-will-face-from-climate-change-1226036.

42. Carolyn Gramling, "Four Ways to Put the 100-Degree Arctic Heat Record in Context," *Science News*, July 1, 2020, https://www.sciencenews.org/article/climate-new-high-temperature-heat-record-arctic-siberia-context; see also "It's Official: July Was the Earth's Hottest Month on Record," NOAA, August 13, 2021, https://www.noaa.gov/news/its-official-july-2021-was-earths-hottest-month-on-record.

43. Gary Ferguson, *Land on Fire: The New Reality of Wildfire in the West* (Portland, OR: Timber Press, 2017); and Michael Kodas, *Megafire: The Race to Extinguish a Deadly Epidemic of Flame* (Boston: Houghton Mifflin Harcourt, 2017).

44. Somini Sengupta, "How Europe Turned into a Perfect Landscape for Wildfires," *New York Times*, February 5, 2020, https://www.nytimes.com/2020/02/05/climate/forests-europe-climate-changed.html.

45. *Our Future on Earth*, 2020, p. 29, https://futureearth.org/publications/our-future-on-earth/.

46. Some geologists argue that the period begins with the Industrial Revolution of the eighteenth and nineteenth centuries, when people started "reforming" land in significant ways and using new technologies such as coal and steam power. Other scientists, such as Simon L. Lewis and Mark A. Maslin, argue that the geological period started much earlier, around 1610, reflecting early capitalist trading systems and the European invasion of the New World, which included the genocide of Indigenous peoples and enslavement of Africans. Despite scientific disagreement about when the Anthropocene started, there is consensus that human impacts on the natural world have been speeding up in

recent decades. See Lewis and Maslin, "Defining the Anthropocene," *Nature* 519 (2015): 171–80.

However, the concept of the Anthropocene has also been criticized. Some critics point to the way the concept prioritizes a European historical narrative that places western (primarily English) capitalism as the primary driver and agent of planetary change. As the anthropologist Kathleen Morrison argues, the concept perpetuates western structures of thinking and "hides a disturbing extension of colonial discourse into a postcolonial world." Indigenous scholars such as Kim Tallbear are also critical of the Anthropocene concept but take a different view, arguing that irrespective of the historical moment being discussed, Indigenous peoples have always worked with the natural world in a relationship of respect, interconnection, and stewardship, not domination and exploitation as is typical of how the west relates to the natural world. See Paul J. Crutzen, "Geology of Mankind: The Anthropocene," *Nature* 415 (2002): 23; Lewis and Maslin, "Defining the Anthropocene"; Gaia Vince, *Adventures in the Anthropocene: A Journey to the Heart of the Planet We Made* (Minneapolis: Milkweed Editions, 2014); Kathleen D. Morrison, "Provincializing the Anthropocene," *Seminar: The Monthly Symposium* 673 (September 2015, Nature and History): 75–80, at 76; Elizabeth Kolbert. *The Sixth Extinction: An Unnatural History* (New York: Picador, 2015); Kim Tallbear, "Caretaking Relations, Not American Dreaming," *Kalfou: A Journal of Comparative and Relational Ethnic Studies* 6, no. 1 (2019): 24–41.

47. Jessica K. Weir, "Lives in Connection," in *Manifesto for Living in the Anthropocene*, ed. Katherine Gibson, Deborah Bird Rose, and Ruth Fincher, 17–21 (Brooklyn, NY: Punctum Books, 2015).

48. Cited in Jessica Corbett, "Scientists Warn of 'Ghastly Future' Unless Policymakers Act Now," *Defend Democracy Now*, June 2, 2021, http://www.defenddemocracy.press/scientists-warn-of-ghastly-future-unless-policymakers-act-now/.

49. This was one of the Trump administration's numerous retreats from international agreements and associations that included pulling out of the Trans-Pacific Partnership Agreement in 2017, the Iran Nuclear Agreement in 2018, and the World Health Organization in 2020.

50. Alfredo Saad-Filho and Lecio Morais, *Brazil: Neoliberalism Versus Democracy* (London: Pluto Press, 2018), 163. See also Alison J. Ayers and Alfredo Saad-Filho, "Democracy Against Neoliberalism: Paradoxes, Limitations, Transcendence," *Critical Sociology* 41, nos. 4–5 (2015): 597–618; and Aurélien Mondon, *The Mainstreaming of the Extreme Right in France and Australia: A Populist Hegemony* (New York: Routledge, 2016).

51. Cited in David Crowe, "Deputy PM Slams People Raising Climate Change in Relation to NSW Bushfires," *Sydney Morning Herald*, November 11, 2019, https://www.smh.com.au/politics/federal/raving -inner-city-lunatics-michael-mccormack-dismisses-link-between-climate -change-and-bushfires-20191111-p539ap.html.

52. See Boaventura de Sousa Santos, *The End of the Cognitive Empire: The Coming of Age of Epistemologies of the South* (Durham, NC: Duke University Press, 2018); and Julie Sze, *Environmental Justice in a Moment of Danger* (Oakland: University of California Press, 2020).

53. Global Greens Charter, accessed January 1, 2021, https://www .globalgreens.org/globalcharter.

54. The anthropologist Mark Schuller highlights this approach by calling for a need to develop a "radical empathy." See Schuller, *Humanity's Last Stand: Confronting Global Catastrophe* (New Brunswick, NJ: Rutgers University Press, 2021).

55. See Vandana Shiva, *Making Peace with the Earth* (London: Pluto Press, 2013); Daniel R. Wildcat, *Red Alert! Saving the Planet with Indigenous Knowledge* (Golden, CO: Fulcrum Publishing, 2009); Soren C. Larsen and Jay T. Johnson, *Being Together in Place: Indigenous Coexistence in a More Than Human World* (Minneapolis: University of Minnesota Press, 2017); and Robin Wall Kimmerer, *Braiding Sweetgrass: Indigenous Wisdom, Scientific Knowledge, and the Teachings of Plants* (Minneapolis: Milkweed Editions, 2013).

56. Writes Naomi Klein, "Just as climate denialism has become a core identity issue on the right, utterly entwined with defending current systems of power and wealth, the scientific reality of climate change must, for progressives, occupy a central place in a coherent narrative about the perils of unrestrained greed and the need for real alternatives." She goes on optimistically: "Building such a transformative movement may not be as hard as it first appears" (Klein, *On Fire*, 77).

57. The human/nature divide is associated with the Cartesian logic of modern western science that emerged with the philosophical writings of René Descartes in seventeenth-century France. This logic separates humans from nature, presenting humans as biologically and morally superior and ultimately able to control nature. This divide was critiqued by Karl Marx and other political thinkers in the nineteenth and twentieth centuries and has come under renewed attack in recent decades. An excellent analysis of ecological concerns being tied to a critique of capitalism can be found in John Bellamy Foster's book *The Return of Nature: Socialism and Ecology* (New York: Monthly Review Press, 2020).

Jason W. Moore is another scholar critical of the human/nature binary. He argues that the binary is problematic for lumping all people into one category, presenting humankind as a homogenous mass acting upon and exploiting nature. Against this abstraction, Moore argues that it was primarily European propertied white men who engaged in early capitalism. And it was these specific men who are historically responsible for profiting from a system of unequal wealth distribution that led to the massive environmental degradation over centuries that has cumulatively changed today's planetary climate. However, the problem goes deeper than certain people holding power and privilege. As Moore points out, the human/nature binary is built on a fallacy, because from the start Europeans deemed certain people (Indigenous and enslaved people) as nonhuman and placed them in the *nature* category. Treating people as objects of property caused their humanity to be cheapened and degraded. This process was necessary to justify their exploitation over many centuries, deny them legal standing and protections, and exclude them from social and political life.

This argument basically says that the human/nature divide was essential for the rise of European capitalism. On its colonial frontiers, colonialists extracted value from nature in the form of material resources (timber and land) and free labor (enslaved Africans, Indigenous peoples, and women). Given this extractive system, Moore argues, *Anthropocene* should be replaced with the term *Capitalocene* because it explicitly references capitalism and its relations of power, violence, and cumulative exploitation impacting the "web of life." To put this differently, the Anthropocene charts the consequences of human activity on the geological record, whereas the Capitalocene seeks to understand the humanity-in-nature coproduction of planetary crisis. I agree with this assessment and think the term "Capitalocene" better reflects how extractive capitalism and the dominance of corporations are shaping our physical landscapes. However, if one follows Cedric Robinson's concept of "racial capitalism," it can be argued that the logics of domination and exploitation of people and nature were evident in feudal times and existed prior to modern forms of capital accumulation. See Foster, *The Return of Nature*; Kohei Saito, *Karl Marx's Ecosocialism: Capitalism, Nature, and the Unfinished Critique of Political Economy* (New York: Monthly Review Press, 2017); Jason W. Moore, *Capitalism in the Web of Life: Ecology and the Accumulation of Capital* (London: Verso, 2015); Cedric Robinson, *Black Marxism: The Making of the Black Radical Tradition* (London: Zed Press, 1983); Hans A. Baer, *Global Capitalism and Climate Change: The Need for an Alternative World System* (Lanham, MD: AltaMira Press, 2012); and Kathryn Yusoff, *A Billion Black Anthropocenes or None* (Minneapolis: University of Minnesota Press, 2018).

58. See William Cronon, *Changes in the Land: Indians, Colonists, and the Ecology of New England* (rev. ed.; New York: Hill and Wang, 2003).

59. Conservation movements that emerged in the nineteenth century in the United States, Canada, Australia, and across Europe are historically deeply racist in their exclusion of Indigenous and minority communities from an idealized "empty" landscape. Says Jon Christensen, an environmental historian, "The [John] Muir ideal of the lone white man at one with nature in the wilderness excludes all kinds of people from that relationship" (cited in Lucy Tompkins, "Sierra Club Says It Must Confront Its Founder's Racism," *New York Times*, July 24, 2020). See also Dorceta E. Taylor, *The Rise of the American Conservation Movement: Power, Privilege, and Environmental Protection* (Durham, NC: Duke University Press, 2016); Mark Dowie, *Conservation Refugees: The Hundred-Year Conflict Between Global Conservation and Native Peoples* (Cambridge, MA: MIT Press, 2011); and Joe Curnow and Anjali Helferty, "Contradictions of Solidarity: Whiteness, Settler Coloniality, and the Mainstream Environmental Movement," *Environment and Society* 9, no. 1 (2018): 145–63.

60. See Donna Haraway, *When Species Meet* (Minneapolis: University of Minnesota Press, 2008); and Helen Merrick, "Naturecultures

and Feminist Materialism," in *Routledge Handbook of Gender and Environment*, ed. Sherilyn MacGregor, 101–14 (New York: Routledge, 2017).

61. In a parallel way, an exciting and innovative body of scholarship calls for us to think through oceans. See Epeli Hau'ofa, "The Ocean in Us," *Contemporary Pacific* 10, no. 2 (1998): 392–410; Karin Amimoto Ingersoll, *Waves of Knowing: A Seascape Epistemology* (Durham, NC: Duke University Press, 2016); Renisa Mawani, *Across Oceans of Law: The* Komagata Maru *and Jurisdiction in the Time of Empire* (Durham, NC: Duke University Press, 2018); Sonja Boon, *What the Oceans Remember: Searching for Belonging and Home* (Waterloo, ON: Wilfrid Laurier University Press, 2019); and Irus Braverman and Elizabeth R. Johnson, eds., *Blue Legalities: The Life and Laws of the Sea* (Durham, NC: Duke University Press, 2020).

62. Timothy Mitchell, *Rule of Experts: Egypt, Techno-Politics, Modernity* (Berkeley: University of California Press, 2002): 19–53; Moore, *Capitalism in the Web of Life*.

63. Faye Brown, "Smoke from Australia's Wildfires Has Travelled over 9,000 Miles to Brazil," *Metro*, January 10, 2020, https://metro .co.uk/2020/01/10/smoke-australias-wildfires-travelled-9000-miles -brazil-12036242/.

64. Global Forest Watch Fires (GFW Fires) is an online platform for monitoring and responding to forest and land fires using near-real-time information. GFW Fires can empower people to better combat harmful fires before they burn out of control and to hold accountable those who may have burned forests illegally. Accessed January 10, 2021, https:// fires.globalforestwatch.org/home/.

65. See Ulrick Beck, *The Metamorphosis of the World: How Climate Change Is Transforming Our Concept of the World* (Cambridge, UK: Polity Press, 2016).

66. Thinking critically about how the historical legacies of capitalism, colonialism, and imperialism inform contemporary global processes, and how these are then experienced in the lives of ordinary people, is one of the central concerns of the new field of global studies. Another central concern is the embracing of plural epistemologies that challenge the dominance of western knowledge and its modes of production. For a discussion of the field of global studies, see Eve Darian-Smith and Philip C. McCarty, *The Global Turn: Theories, Research Designs, and Methods for*

Global Studies (Oakland: University of California Press, 2017); and Manfred Steger and Amentahru Wahlrab, *What Is Global Studies? Theory and Practice* (London: Routledge, 2017).

67. Donna J. Haraway, *Staying with the Trouble: Making Kin in the Chthulucene* (Durham, NC: Duke University Press, 2016). See also Zoe Todd, "Refracting the State Through Human-Fish Relations: Fishing, Indigenous Legal Orders, and Colonialism in North/Western Canada," *DIES: Decolonization, Indigeneity, Education, and Society* 7, no. 1 (2018): 60–75.

Chapter 2: Fire as Profit

1. Wendy Brown, *Undoing the Demos: Neoliberalism's Stealth Revolution* (London: Zone Books, 2015), 4.

2. See Zephyr Teachout, *Break 'Em Up: Recovering Our Freedom from Big Ag, Big Tech, and Big Business* (New York: All Points Books, 2020).

3. CBS San Francisco, "Families of Camp Fire Victims Call PG&E 'Mass Murderers . . . Thieves, Liars,'" June 17, 2020, https://sanfran cisco.cbslocal.com/2020/06/17/families-camp-fire-victims-call-pge -mass-murderers-thieves-liars/.

4. See Gerald Horne, *The Dawning of the Apocalypse: The Roots of Slavery, White Supremacy, Settler Colonialism, and Capitalism in the Long Sixteenth Century* (New York: Monthly Review Press, 2020).

5. *Neoliberalism* is a complex and much-debated term, but few disagree that it embraces the pursuit of market efficiencies and profit as its primary objectives. This has resulted in a worldwide attack on the principles of liberal democracy that include political, social, and legal accountability, as well as equity in the distribution of resources. See Noam Chomsky, *Profit over People: Neoliberalism and Global Order* (New York: Seven Stories Press, 1999); David Harvey, *A Brief History of Neoliberalism* (Oxford: Oxford University Press, 2007); Philip Mirowski and Dieter Plehwe, *The Road from Mount Pèlerin: The Making of the Neoliberal Thought Collective* (Cambridge, MA: Harvard University Press, 2009); and Quinn Slobodian, *Globalists: The End of Empire and the Birth of Neoliberalism* (Cambridge, MA: Harvard University Press, 2018).

6. Fassil Demissie, ed., *Land Grabbing in Africa: The Race for Africa's Rich Farmland* (New York: Routledge, 2017); Stefano Liberti, *Land Grabbing: Journeys in the New Colonialism*, trans. Enda Flannelly (London: Verso, 2013); Jason Hickel, *The Divide: Global Inequality from Conquest to Free Markets* (New York: W.W. Norton, 2018).

7. Niels M. Myers and Jorgen S. Nørgård, "Policy Means for Sustainable Energy Scenarios," in "Collection of Extended Abstracts," *Proceedings of the International Conference on Energy, Environment, and Health—Optimisation of Future Energy Systems*, May 31–June 2, 2010, Carlsberg Academy, Copenhagen, Denmark, 133–37, at 134, https://web.archive.org/web/20161009062231/http://www2.dmu.dk/1_om_dmu/2_afdelinger/3_atmi/ceeh/Collection%20of%20Extended%20Abstracts.pdf#page=134.

8. Global inequality was an acute problem well before the global COVID-19 pandemic. UN analysts now predict the consequences of the pandemic could push half a billion people into poverty—a staggering 8 percent of the total human population. See United Nations, "COVID-19 Could See over 200 Million More Pushed into Extreme Poverty, New UN Development Report Finds," December 3, 2020, https://news.un.org/en/story/2020/12/1079152.

9. It is not a coincidence that in September 2020, as California was suffering under long-term drought and what were then the most devastating wildfires in the state's history, water started to be traded on Wall Street, along with other commodities such as oil, gold, and minerals. "Farmers, hedge funds, and municipalities are now able to hedge against—or bet on—future water availability in California, the biggest US agriculture market and world's fifth-largest economy." The water futures contracts are meant to protect against water-price fluctuations for big water-consuming companies that include electricity utilities and almond producers, as well as serve as a "scarcity gauge for investors worldwide." See Kim Chipman, "California Water Futures Begin Trading amid Fear of Scarcity," *Bloomberg Green*, December 6, 2020, https://www.bloomberg.com/news/articles/2020–12–06/water-futures-to-start-trading-amid-growing-fears-of-scarcity.

10. Large-scale land grabbing has been going on for centuries. As Megan Black reminds us in her original and compelling book *The Global*

Interior: Mineral Frontiers and American Power (Cambridge, MA: Harvard University Press, 2018), land grabbing and extractivism in overseas territories has long been part of the United States' covert expansionist strategy to build an empire and global hegemonic power. The country's pursuit of minerals and related management of the environment provided a means and rationale for overseas intervention justified by the "innocent" search for natural resources, delinked from the politics of imperial aggression.

11. GRAIN, "Seized: The 2008 Landgrab for Food and Financial Security," October 24, 2008, https://www.grain.org/article/entries/93-seized-the-2008-landgrab-for-food-and-financial-security.pd.

12. Saturnino M. Borras et al., *Land Grabbing and Human Rights: The Involvement of European Corporate and Financial Entities in Land Grabbing Outside the European Union*, European Parliament's Subcommittee on Human Rights, 2016, https://www.europarl.europa.eu/RegData/etudes/STUD/2016/578007/EXPO_STU(2016)578007_EN.pdf.

13. Saskia Sassen, *Expulsions: Brutality and Complexity in the Global Economy* (Cambridge, MA: Belknap, 2014), 10; Eve Darian-Smith, "Who Owns the World? Landscapes of Sovereignty, Property, Dispossession," *Journal of the Oxford Centre for Socio-Legal Studies* 1, no. 2 (2016), https://www.law.ox.ac.uk/content/annual-socio-legal-lecture-who-owns-world-landscapes-sovereignty-property-dispossession.

14. Nick Estes, "Bill Gates Is the Biggest Private Owner of Farmland in the United States. Why?," *Guardian*, April 5, 2021, https://www.theguardian.com/commentisfree/2021/apr/05/bill-gates-climate-crisis-farmland.

15. Rachel Carson, *Silent Spring* (40th anniv. ed.; Boston: Houghton Mifflin, 2002); Joni Seager, *Carson's "Silent Spring": A Reader's Guide* (London: Bloomsbury, 2014); Adam Rome, *The Genius of Earth Day: How a 1970 Teach-In Unexpectedly Made the First Green Generation* (New York: Hill and Wang, 2014).

16. Naomi Oreskes and Erick M. Conway, *Merchants of Doubt: How a Handful of Scientists Obscured the Truth on Issues from Tobacco Smoke to Global Warming* (London: Bloomsbury Press, 2010), 216–39.

17 Kolbert, *The Sixth Extinction.*

18. Christian Parenti, "'The Limits to Growth': A Book That Launched a Movement," *Nation*, December 5, 2012, https://www.the nation.com/article/archive/limits-growth-book-launched-movement/.

19. Jason Hickel, *Less Is More: How Degrowth Will Save the World* (London: William Heinemann, 2020).

20. Vandana Shiva, *Making Peace with the Earth* (London: Pluto Press, 2013), 3.

21. Naomi Klein, *This Changes Everything: Capitalism vs. the Climate* (New York: Simon & Schuster Paperbacks, 2014), 169.

22. Klein, *This Changes Everything*, 169.

23. Steve Lerner, *Sacrifice Zones: The Front Lines of Toxic Chemical Exposure in the United States* (Cambridge, MA: MIT Press, 2010).

24. Charles M. Coleman, *P.G. and E. of California: The Centennial Story of Pacific Gas and Electric Company, 1852–1952* (New York: McGraw-Hill, 1952).

25. PG&E, "Decommissioning Diablo Canyon Power Plant in 2025," https://www.pge.com/en_US/safety/how-the-system-works/diablo -canyon-power-plant/diablo-canyon-power-plant/diablo-decommission ing.page.

26. Katie Worth, Karin Pinchin, and Lucie Sullivan, "'Deflect, Delay, Defer': Decade of Pacific Gas & Electric Wildfire Safety Pushback Preceded Disasters," Frontline, August 18, 2020, https://www.pbs.org /wgbh/frontline/article/pge-california-wildfire-safety-pushback/.

27. Ivan Penn and Peter Eavis, "PG&E Says It's Guilty in 84 Deaths from Fire," *New York Times*, June 16, 2020, https://www.nytimes.com /2020/06/16/business/energy-environment/pge-camp-fire-california -wildfires.html; Ivan Penn, "PG&E Is Ordered to Pay $3.5 Million Fine for Causing Deadly Wildfire," *New York Times*, June 18, 2020, https:// www.nytimes.com/2020/06/18/business/energy-environment/pge -camp-fire-sentenced.html.

28. Russell Gold, "PG&E: The First Climate-Change Bankruptcy, Probably Not the Last," *Wall Street Journal*, January 18, 2019, https:// www.wsj.com/articles/pg-e-wildfires-and-the-first-climate-change -bankruptcy-11547820006.

29. Rebecca Leber, "PG&E Was Once Part of the Climate-Denial Machine That Helped Fuel California's Blackout Crisis," *Mother Jones*,

October 11, 2019, https://www.motherjones.com/environment/2019/10/pge-blackouts-climate-denial/.

30. Denial of scientific evidence has a long history in the United States and is commonly associated with the antiscience campaigns sponsored by the tobacco industry against medical proof that smoking causes cancer. See Oreskes and Conway, *Merchants of Doubt*; and Barbara Freese, *Industrial-Strength Denial: Eight Stories of Corporations Defending the Indefensible, from the Slave Trade to Climate Change* (Oakland: University of California Press, 2020).

31. Suzanne Goldenberg, "Exxon Knew of Climate Change in 1981, Email Says—but It Funded Deniers for 27 More Years," *Guardian*, July 8, 2015, https://www.theguardian.com/environment/2015/jul/08/exxon-climate-change-1981-climate-denier-funding.

32. Riley E. Dunlap and Peter J. Jacques, "Climate Change Denial Books and Conservative Think Tanks: Exploring the Connection," *American Behavioral Scientist* 57, no. 6 (2013): 699–731; Clive Hamilton, *Requiem for a Species: Why We Resist the Truth About Climate Change* (New York: Routledge, 2010); Michael Mann and Tom Toles, *The Madhouse Effect: How Climate Change Denial Is Threatening Our Planet, Destroying Our Politics, and Driving Us Crazy* (New York: Columbia University Press, 2018); Michael Mann, *The New Climate War: The Fight to Take Back Our Planet* (New York: Public Affairs, 2021).

33. John Vidal, "Climate Sceptic Willie Soon Received $1m from Oil Companies, Papers Show," *Guardian*, June 28, 2011, https://www.theguardian.com/environment/2011/jun/28/climate-change-sceptic-willie-soon.

34. Bruno Latour, *Down to Earth: Politics in the New Climatic Regime* (Cambridge, UK: Polity, 2017), 19. The italics are Latour's.

35. Michael Wolff, *The Man Who Owns the News: Inside the Secret World of Rupert Murdoch* (New York: Broadway Books, 2010).

36. Oreskes and Conway, *Merchants of Doubt*, 214.

37. Andy Gregory, "Murdoch Media Empire a 'Cancer on Democracy,' Former Australian Prime Minister Says," *Independent*, October 12, 2020, https://www.independent.co.uk/news/world/australasia/rupert-murdoch-kevin-rudd-petition-newscorp-australia-james-b989607.html.

38. Ivan Penn, "'This Did Not Go Well': Inside PG&E's Blackout Control Room," *New York Times*, October 12, 2019, https://www.nytimes.com/2019/10/12/business/pge-california-outage.html; Brandon Rittiman, "'Blood Money': California Politicians and Campaigns Received $2.1 Million from Bankrupt, Guilty PG&E," ABC News, March 2, 2021, https://www.abc10.com/article/news/politics/california-politicians-campaigns-2-million-pge/103-4161feb9-1591-4ffc-9fda-d9c49d7173b8.

39. Ian W. McLean, *Why Australia Prospered: The Shifting Sources of Economic Growth* (Princeton, NJ: Princeton University Press, 2016).

40. Geoffrey Morrison, "Take a Tour of the Super Pit: The Largest Open-Pit Gold Mine in Australia," CNET, May 4, 2015, https://www.cnet.com/news/take-a-tour-of-the-super-pit-the-largest-open-pit-gold-mine-in-australia/.

41. "The Future Beyond the Pit," Government of Western Australia report posted on the Western Australian Museum website, accessed January 15, 2021, http://museum.wa.gov.au/explore/wa-goldfields/environmental-impacts/future-beyond-pit.

42. Charles Roche and Simon Judd, *Ground Truths: Taking Responsibility for Australia's Mining Legacies* (n.p.: Mineral Policy Institute, 2016), 5, http://www.mpi.org.au/wp-content/uploads/2016/06/Ground-Truths-2016-web.pdf.

43. Malcolm Knox, *Boom: The Underground History of Australia, from Gold Rush to GFC* (Melbourne: Viking, 2013).

44. In 2017 the Palmer United Party was deregistered and then revived a year later and reregistered under the name United Australia Party. As of 2021, it is led by Craig Kelly, who holds a seat in the Australian House of Representatives.

45. For instance, Brendan Pearson, former CEO of Australia's peak mining lobby group, the Minerals Council of Australia, joined Prime Minister Scott Morrison's conservative government as a senior advisor in 2019 and has worked hard to defend corporate mining interests. See Amy Remeikis, "Adviser Whose Former Lobby Group Provided Lump of Coal to Scott Morrison Made Ambassador to OECD," *Guardian*, September 17, 2021, https://www.theguardian.com/business/2021/sep/17/adviser-whose-former-lobby-group-provided-lump-of-coal-to-scott-morrison-made-ambassador-to-oecd.

46. Cited in Christopher Knaus, "Mining Firms Worked to Kill Off Climate Action in Australia, Says Ex-PM," *Guardian*, October 10, 2019, https://www.theguardian.com/environment/2019/oct/10/mining-firms-worked-kill-off-climate-action-australia-ex-pm-kevin-rudd.

47. Cited in Knaus, "Mining Firms Worked to Kill Off Climate Action."

48. Clive Hamilton, *Silencing Dissent: How the Australian Government Is Controlling Public Opinion and Stifling Debate* (Melbourne: Allen & Unwin, 2007).

49. Sora Park et al., "Digital News Report: Australia 2020," News and Media Research Centre, University of Canberra, https://apo.org.au/node/305057.

50. Elle Hunt, "Scott Morrison and Ray Hadley Laugh About Coal Prop: 'Great Stunt,'" *Guardian*, February 12, 2017, https://www.theguardian.com/australia-news/2017/feb/13/scott-morrison-and-ray-hadley-laugh-about-coal-prop-great-stunt.

51. Jeff Goodall, "The World's Most Insane Energy Project Moves Ahead," *Rolling Stone*, June 14, 2019, https://www.rollingstone.com/politics/politics-news/adani-mine-australia-climate-change-848315/.

52. "David Attenborough Slams Australian PM on Climate Record," ABC News, September 24, 2019, https://www.abc.net.au/triplej/programs/hack/sir-david-attenborough-slams-scott-morrison-on-climate-record/11533566.

53. Anne Layton-Bennett, "Dear Mr. Morrison," in *From the Ashes: A Poetry Collection in Support of the 2019–2020 Australian Bushfire Relief Effort*, ed. C.S. Huges, 61 (Korumburra, Australia: Maximum Felix Media, 2020).

54. Perry Anderson, *Brazil Apart: 1964–2019* (London: Verso, 2019), 180–85.

55. David Price, *Before the Bulldozer: The Nambiquara Indians and the World Bank* (Washington, DC: Seven Locks Press, 1989); Chris Feliciano Arnold, *The Third Bank of the River: Power and Survival in the Twenty-First Century Amazon* (New York: Picador, 2018); Jeffrey Hoelle, *Rainforest Cowboys: The Rise of Ranching and Cattle Culture in Western Amazonia* (Austin: University of Texas Press, 2015).

56. See Doug Boucher, Sarah Roquemore, and Estrellita Fitzhugh, "Brazil's Success in Reducing Deforestation," *Tropical Conservation*

Science 6, no. 3 (2013): 426–45; World Bank, "Brazil's INDC Restoration and Reforestation Target: Analysis of INDC Land-Use Targets," Washington, DC, 2017, https://openknowledge.worldbank.org/handle /10986/28588.

57. Lisa Viscidi and Nate Graham, "Brazil Was a Global Leader on Climate Change. Now It's a Threat," *Foreign Policy*, January 4, 2019, https://foreignpolicy.com/2019/01/04/brazil-was-a-global-leader-on -climate-change-now-its-a-threat/.

58. "As a crop defined largely by the value and usefulness of its co-products—namely, soybean meal and soybean oil–soy might be regarded as a fundamentally flexible crop. Of the world's total soy production, only 6 percent is consumed in the form of whole beans, tofu, or other whole-soy and fermented foods. The remaining 94 percent is crushed, either mechanically or chemically, to produce soybean meal and oil for further processing. . . . Worldwide, the meal portion of the crush is predominantly used in livestock feed (98 percent), while the remainder becomes soy flour and soy protein for food processing industries. Soy oil is largely refined as edible oil (95 percent), with the rest funneled to industrial products, including biodiesel." See Gustavo de L.T. Oliveira and Mindi Schneider, "The Politics of Flexing Soybeans: China, Brazil, and Global Agroindustrial Restructuring," *Journal of Peasant Studies* 43, no. 1 (2016): 167–94, at 168.

59. Jake Spring, "Soy Boom Destroys Brazil's Tropical Savanna: A Reuters Special Report," Reuters Investigates, August 28, 2018, https:// www.reuters.com/investigates/special-report/brazil-deforestation/.

60. Spring, "Soy Boom Destroys Brazil's Tropical Savanna."

61. Cited in Jake Spring, "Brazil Minister Calls for Environmental Deregulation While Public Distracted by COVID," Reuters, May 22, 2020, https://www.reuters.com/article/us-brazil-politics-environment /brazil-minister-calls-for-environmental-deregulation-while-public -distracted-by-covid-idUSKBN22Y30Y.

62. "Deforestation on Amazonian Public Lands Takes Off and Could Fuel Fire Season," IPAM Amazônia, April 24, 2020, https:// ipam.org.br/deforestation-on-amazonian-public-lands-takes-off-and -could-fuel-fire-season/.

63. Technical Note, IPAM Amazônia, April 2020, no. 3, https:// ipam.org.br/wp-content/uploads/2020/04/NT3-Fire-2019.pdf.

64. Cited in Joao Fellet and Charlotte Pamment, "Amazon Rainforest Plots Sold via Facebook Marketplace Ads," BBC News, February 12, 2021, https://www.bbc.com/news/technology-56168844.

65. Cited in Erik Ortiz, "How the Amazon's Fires, Deforestation Affect the US Midwest," NBC News, August 24, 2019, https://www.nbcnews.com/news/world/how-amazon-s-fires-deforestation-affect-u-s-midwest-n1045886.

66. Cited in Anna Jean Kaiser, "Is the Amazon Really 'the Lungs' of Planet Earth? No, It's More Like Our Sink," *Chicago Tribune*, August 27, 2019, https://www.chicagotribune.com/nation-world/ct-nw-cb-amazon-fires-explainer-20190827-oneqsy6c6nfuxb33bictnewnbm-story.html.

67. David Waltner-Toews, *On Pandemics: Deadly Diseases from Bubonic Plague to Coronavirus* (Vancouver, BC: Greystone Books, 2020).

68. Cited in *Preventing the Next Pandemic—Zoonotic Diseases and How to Break the Chain of Transmission*, UN Environment Programme (UNEP) Report, July 6, 2020, https://www.unenvironment.org/resources/report/preventing-future-zoonotic-disease-outbreaks-protecting-environment-animals-and.

Chapter 3: Fire as Weapon

1. Ivan Krastev, "Biden Can't Decide What Counts as a 'Democracy,'" *New York Times*, May 12, 2021, https://www.nytimes.com/2021/05/12/opinion/biden-democracy-alliance.html.

2. See Bernhard Forchtner, ed., *The Far Right and the Environment: Politics, Discourse, and Communication* (New York: Routledge, 2020). See also proceedings from the Political Ecologies of the Far Right conference, held at Lund University in November 2019, and related subsequent events, https://undisciplinedenvironments.org/2020/05/13/political-ecologies-of-the-far-right-an-introduction-to-the-series/. There is also emerging a body of scholarship examining far-right populism and antienvironmentalism specifically. Within this scholarship there is consensus that far-right populists are predominantly antienvironment and tend to be climate deniers or skeptics. However, there is debate about the reasons. Some scholars weigh the respective relevance of ideologies held by right-wing populists (that is, ultranationalist, nativist, antielitist) and the structural conditions that have created

NOTES TO CHAPTER 3   167

socioeconomic disadvantage in determining populists' antienvironmental perspectives. Other scholars show, not surprisingly, that extreme-right political parties are adept at manipulating and exploiting the populist right for their own political gain. This helps us understand why someone like Brazil's president Jair Bolsonaro can openly encourage the burning of the Amazonian rainforest and concurrently fortify the country's solar and wind energy capacities. See Matthew Lockwood, "Right-Wing Populism and the Climate Change Agenda: Exploring Linkages," *Environmental Politics* 27, no. 4 (2018): 712–32; Robert A. Huber, "The Role of Populist Attitudes in Explaining Climate Change Skepticism and Support for Environmental Protection," *Environmental Politics* 29, no. 6 (2020): 959–82; Jale Tosun and Marc Debus, "Right-Wing Populist Parties and Environmental Politics: Insights from the Austrian Freedom Party's Support for the Glyphosate Ban," *Environmental Politics* 30, nos. 1–2 (2021): 224–44.

3. The concept of authoritarianism varies across different societies and historical times and has been the subject of a vast body of scholarship. Much of the literature focuses on the authoritarian regimes associated with World War II. But more recently commentators have been exploring the historical legacies of fascism in connection with extreme-right social movements and new forms of repressive government in the twenty-first century. See Juan Linz, *Totalitarian and Authoritarian Regimes* (Boulder, CO: Lynne Rienner Publishers, 2000); Berch Berberoglu, *The Global Rise of Authoritarianism in the 21st Century: Crisis of Neoliberal Globalization and the Nationalist Response* (London: Routledge, 2020); Patrik Hermansson, David Lawrence, Joe Mulhall, and Simon Murdoch, *The International Alt-Right: Fascism for the 21st Century?* (New York: Routledge, 2020); and Theo Horesh, *The Fascism This Time: And the Global Future of Democracy* (Boulder, CO: Cosmopolis Press, 2020).

4. See Wendy Brown, *In the Ruins of Neoliberalism: The Rise of Antidemocratic Politics in the West* (New York: Columbia University Press, 2019); and Anne Applebaum, *Twilight of Democracy: The Seductive Lure of Authoritarianism* (New York: Doubleday, 2020).

5. See Ian Bruff, "The Rise of Authoritarian Neoliberalism," *Rethinking Marxism* 26, no. 1 (2014): 113–29; Bonn Juego, "Authoritarian Neoliberalism: Its Ideological Antecedents and Policy Manifestations

from Carl Schmitt's Political Economy of Governance," *Administrative Culture* 19, no. 1 (2018): 105–36; Ayers and Saad-Filho, "Democracy Against Neoliberalism."

6. For example, the International Commission for the Conservation of Atlantic Tunas was formally established in 1966 to protect fish populations, but over decades was redirected by neoliberal priorities and in practice facilitated overfishing and maximizing countries' economic profits. This led to the methodical and predictable extinction of the giant bluefin tuna. See Jennifer E. Telesca, *Red Gold: The Managed Extinction of the Giant Bluefin Tuna* (Minneapolis: University of Minnesota Press, 2020). For a more general discussion, see Quinn Slobodian, *Globalists: The End of Empire and the Birth of Neoliberalism* (Cambridge, MA: Harvard University Press, 2018).

7. Slobodian, *Globalists: The End of Empire*. See also Nancy MacLean, *Democracy in Chains: The Radical Right's Stealth Plan for America* (New York: Penguin Books, 2018); and Jane Mayer, *Dark Money: The Hidden History of the Billionaires Behind the Rise of the Radical Right* (New York: Penguin, 2017).

8. Jamie Peck reminds us that neoliberalism is not a static idea but in practice morphed over time and between different countries, "made and remade, as a constructed project." See *Constructions of Neoliberal Reason* (Oxford: Oxford University Press, 2010), xii–xiii.

9. Brown, *In the Ruins of Neoliberalism*, 61–62. See also Nicholas Shaxson, *The Finance Curse: How Global Finance Is Making Us All Poorer* (New York: Grove Atlantic, 2019).

10. Jason Hickel, *The Divide: Global Inequality from Conquest to Free Markets* (New York: W.W. Norton, 2018); Eve Darian-Smith, "Global Law Firms in Real-World Contexts: Practical Limitations and Ethical Implications," *Beijing Law Review* 6 (2015): 92–101; Terence C. Halliday and Gregory Shaffer, eds., *Transnational Legal Orders* (Cambridge: Cambridge University Press, 2016); Peer Zumbansen, ed., *The Oxford Handbook of Transnational Law* (Oxford: Oxford University Press, 2021).

11. In 2016, and then again in 2021, approximately 12 million documents were leaked to the international press showing the financial offshore dealings of world leaders, billionaires, and corporate executives.

These bundles of documents are called the Panama Papers and the Pandora Papers, respectively, and reveal the scale of global crime, corruption, and money laundering through deliberately complex offshore company schemes. See Michael S. Schmidt and Steven Lee Myers, "Panama Law Firm's Leaked Files Detail Offshore Accounts Tied to World Leaders," *New York Times*, April 3, 2016, https://www.nytimes.com/2016 /04/04/us/politics/leaked-documents-offshore-accounts-putin.html; and "Pandora Papers: An Offshore Data Tsunami," International Consortium of Investigative Journalists, October 3, 2021, https://www .icij.org/investigations/pandora-papers/about-pandora-papers-leak -dataset/.

12. Cemal Burak Tansel, ed., *States of Discipline: Authoritarian Neoliberalism and the Contested Reproduction of Capitalist Order* (London: Rowman & Littlefield, 2017), 13.

13. Joseph E. Stiglitz, *Globalization and Its Discontents Revisited: Anti-Globalization in the Era of Trump* (New York: W.W. Norton, 2017).

14. Henry A. Giroux, *The Terror of Neoliberalism: Authoritarianism and the Eclipse of Democracy* (Boulder, CO: Paradigm, 2004.

15. Lisa Duggan, *The Twilight of Equality? Neoliberalism, Cultural Politics, and the Attack on Democracy* (Boston: Beacon Press, 2003).

16. Stephen Metcalf, "Neoliberalism: The Idea That Swallowed the World," *Guardian*, August 18, 2017, https://www.theguardian.com /news/2017/aug/18/neoliberalism-the-idea-that-changed-the-world.

17. Walden Bello, *Counterrevolution: The Global Rise of the Far Right* (Black Point, NS: Fernwood Publishing, 2020); Jens Rydgren, ed., *The Oxford Handbook of the Radical Right* (Oxford: Oxford University Press, 2018).

18. National Democratic Institute (NDI), "A Call to Defend Democracy," June 25, 2020, https://www.ndi.org/publications/call-de fend-democracy.

19. Anatoly Kurmanaev, "Already Frail, Latin American Democracy Faces Wider Threats in Pandemic," *New York Times*, July 30, 2020.

20. Even Steven J. Calabresi, cofounder of the conservative Federalist Society and a committed Republican, wrote that Trump's tweet about possibly postponing the November 2020 election "is fascistic and is itself grounds for the president's immediate impeachment again by the House

of Representatives and removal by the Senate" (in "Trump Might Try to Postpone the Election. That's Unconstitutional," *New York Times*, July 30, 2020, https://www.nytimes.com/2020/07/30/opinion/trump-delay-election-coronavirus.html. Given the insurrection and takeover of the White House encouraged by Trump on January 6, 2021, these comments only a few months earlier seem very prescient.

21. Cited in Peter Baker, "More Than Just a Tweet: Trump's Campaign to Undercut Democracy," *New York Times*, July 31, 2020, https://www.nytimes.com/2020/07/31/us/politics/trump-tweet-democracy.html.

22. Steven Levitsky and Daniel Ziblatt, *How Democracies Die* (New York: Crown, 2018).

23. Boaventura de Sousa Santos, "Trump Won't Take Cyanide," *Critical Legal Thinking*, January 11, 2021, https://criticallegalthinking.com/2021/01/11/trump-wont-take-cyanide/.

24. Christopher R. Browning, "The Suffocation of Democracy," *New York Review of Books*, October 25, 2018, https://www.nybooks.com/articles/2018/10/25/suffocation-of-democracy/.

25. See Vijay Prashad, ed., *Strongmen: Trump-Modi-Erdogan-Duterte-Putin* (New Delhi: LeftWord, 2018).

26. Javier Corrales, "The Authoritarian Resurgence: Autocratic Legalism in Venezuela," *Journal of Democracy* 26, no. 2 (2015): 37–51, at 38; and Kim L. Scheppele, "Autocratic Legalism," *University of Chicago Law Review* 85, no. 2 (2018): 545–83, at 571 and 582.

27. Sarah Repucci, *Freedom in the World 2020: A Leaderless Struggle for Democracy*, Freedom House Report, 2020, p. 9, https://freedomhouse.org/sites/default/files/2020-02/FIW_2020_REPORT_BOOKLET_Final.pdf.

28. William E. Connolly, *Aspirational Fascism: The Struggle for Multifaceted Democracy Under Trumpism* (Minneapolis: University of Minnesota Press, 2017).

29. Stuart Stevens, "I Hope This Is Not Another Lie About the Republican Party," *New York Times*, July 29, 2020, https://www.nytimes.com/2020/07/29/opinion/trump-republican-party-racism.html.

30. Browning, "The Suffocation of Democracy"; see also Jack Jackson, *Law Without Future: Anti-Constitutional Politics and the American Right* (Philadelphia: University of Pennsylvania Press, 2019).

31. Robert A. Pape and Keven Ruby, "The Capitol Rioters Aren't Like Other Extremists," *Atlantic*, February 2, 2021.

32. John B. Judis, *The Nationalist Revival: Trade, Immigration, and the Revolt Against Globalization* (New York: Columbia Global Reports, 2018).

33. See Timothy Mitchell, *Carbon Democracy: Political Power in the Age of Oil* (London: Verso, 2013).

34. Arlie Russell Hochschild, *Strangers in Their Own Land: Anger and Mourning on the American Right* (New York: New Press, 2018).

35. Scholars such as Laura Pulido have called Trump's racist outpourings in tweets and public rallies "spectacular racism," underscoring its performative quality and emotional rhetoric designed to rouse and impassion crowds. She and her colleagues argue, "Spectacular racism fuels and coexists with other manifestations of racism, including institutional racism and white privilege. It functions to enhance authoritarian and populist power by solidifying and empowering a political base that is partially animated by white supremacy and xenophobia. It generates loyalty to an individual, rather than to political ideas or institutions. Spectacular racism, white nationalism, and authoritarianism are all distinct but work together in the Trump era." See Laura Pulido, Tianna Bruno, Cristina Falver-Serna, and Cassandra Galentine, "Environmental Deregulation, Spectacular Racism, and White Nationalism in the Trump Era," *Annals of the American Association of Geographers* 109, no. 2 (2019): 520–32.

36. Erik Kojola, "Bringing Back the Mines and a Way of Life: Populism and the Politics of Extraction," *Annals of the American Association of Geographers* 109, no. 2 (2019): 371–81, at 373.

37. Rebecca R. Scott, *Removing Mountains: Extracting Nature and Identity from the Appalachian Coalfields* (Minneapolis: University of Minnesota Press, 2010).

38. Pulido et al., "Environmental Deregulation," 526.

39. Cited in Nadja Popovich, Livia Albeck-Ripka, and Kendra Pierre-Louis, "The Trump Administration Rolled Back More Than 100 Environmental Rules. Here's the Full List," *New York Times*, updated January 20, 2021, https://www.nytimes.com/interactive/2020/climate/trump-environment-rollbacks-list.html.

40. Elizabeth Southerland, "We Can't Let Trump Roll Back 50 Years of Environmental Progress," *Guardian*, April 22, 2020, https://www.theguardian.com/commentisfree/2020/apr/22/earth-day-50-years-anniversary-environment-trump.

41. Justine Calma, "How Scientists Scrambled to Stop Donald Trump's RPA from Wiping Out Climate Data," Verge, March 8, 2021, https://www.theverge.com/22313763/scientists-climate-change-data-rescue-donald-trump; Jeff Brady, "Trump Waives Environmental Reviews, Citing Pandemic Economic Emergency," NPR, June 2, 2020, https://www.npr.org/2020/06/04/870098279/trump-waives-environmental-reviews-citing-pandemic-economic-emergency.

42. Cited in Sharon J. Riley, "Alberta Suspends at Least 19 Monitoring Requirements in Oilsands, Citing Coronavirus Concerns," *Narwhal*, May 5, 2020, https://thenarwhal.ca/alberta-suspends-19-oilsands-environmental-monitoring-requirements-coronavirus-concerns/.

43. The *Climate Transparency Report 2020*, an annual international document detailing the preparations by G20 countries to meet their obligations under the Paris Agreement, lists Australia at the bottom of the pack for its lack of domestic policy, continuing reliance on fossil fuels, and failure to curb rising emissions; https://www.climate-transparency.org/g20-climate-performance/the-climate-transparency-report-2020.

44. The study by scientists from several Australian universities said 2.46 billion reptiles, 180 million birds, 143 million mammals, and 51 million frogs were harmed in the blazes that ripped through the country. See Michael Slezak, "3 Billion Animals Killed or Displaced in Black Summer Bushfires, Study Estimates," ABC News, July 27, 2020, https://www.abc.net.au/news/2020-07-28/3-billion-animals-killed-displaced-in-fires-wwf-study/12497976.

45. Cited in Adam Morton, "Letter by 240 Leading Scientists Calls on Scott Morrison to Stem Extinction Crisis," *Guardian*, October 27, 2019, https://www.theguardian.com/environment/2019/oct/28/toughen-environmental-laws-to-stem-extinction-crisis-scientists-tell-morrison.

46. Seth Borenstein, "AP Fact Check: Trump's Wildfire Tweets Not Grounded in Facts," AP News, November 5, 2019, https://apnews.com/article/ap-fact-check-fires-donald-trump-us-news-forests

-2ceb7a36c7ca4b65b92cb4442cb27f02; Clark Mindock, "Trump Orders FEMA to Send 'No More Money' to California for Forest Fires in Misspelled Tweet," *Independent*, January 9, 2019, https://www.in dependent.co.uk/news/world/americas/us-politics/trump-fema-cali fornia-wildfires-twitter-us-president-federal-money-state-government -shutdown-a8719346.html; John T. Bennett, "'It'll Start Getting Cooler. You Just Watch': Trump States Categorical Denial of Climate Crisis During California Wildfire Visit," *Independent*, September 15, 2020, https://www.independent.co.uk/news/world/americas /us-politics/trump-california-wildfires-climate-crisis-denial-b439267 .html.

47. Joshua Yeager, "Trump Promises More Water for Southern Central Valley Farmers," Recordnet News, February 19, 2020, https:// www.recordnet.com/news/20200219/trump-promises-more-water-for -southern-central-valley-farmers; Kevin Liptak and Gregory Wallace, "Trump Revokes Waiver for California to Set Higher Auto Emissions Standards," CNN, September 18, 2019, https://www.cnn.com/2019 /09/18/politics/epa-trump-california/index.html.

48. Pilita Clark and Lauren Leatherby, "Trump Transcript on Paris Climate Deal Exit—Annotated," *Financial Times*, June 1, 2017, https:// ig.ft.com/trump-paris-agreement-speech-annotator/.

49. Christina Maza, "Far-Right Climate Denial Is Growing in Europe," *New Republic*, November 11, 2019; Olivia Rosane, "German Far Right Party Attacks 16-Year-Old Greta Thunberg Ahead of EU Elections," EcoWatch, May 15, 2019, https://www.ecowatch.com/german-far -right-party-greta-thunberg-2637190985.html; Yasmeen Serhan, "When the Far Right Picks Fights with a Teen," *Atlantic*, August 14, 2021, https://www.theatlantic.com/international/archive/2021/08/greta -thunberg-far-right-climate/619748/.

50. Maza, "Far-Right Climate Denial"; see also Andrew Hoffman, *How Culture Shapes the Climate Change Debate* (Stanford, CA: Stanford University Press, 2015).

51. Paul Krugman, *Arguing with Zombies: Economics, Politics, and the Fight for a Better Future* (New York: W.W. Norton, 2020), 333; see also Naomi Klein, *On Fire: The (Burning) Case for a Green New Deal* (New York: Allen Lane, 2019).

52. Arnim Scheidel et al., "Environmental Conflicts and Defenders: A Global Overview," *Global Environmental Change* 63 (2020): 102–4. See also Nina Likhani, *Who Killed Berta Caceres? Dams, Death Squads, and an Indigenous Defender's Battle for the Planet* (London: Verso, 2020).

53. Frank Uekoetter, *The Green and the Brown: A History of Conservation in Nazi Germany* (Cambridge: Cambridge University Press, 2006); Janet Biehl and Peter Staudenmaier, *Ecofascism Revisited: Lessons from the German Experience* (2nd ed.; Porsgrunn, Norway: New Compass Press, 2011).

54. Hermansson et al., *The International Alt-Right: Fascism for the 21st Century?* (New York: Routledge, 2020).

55. Joel Achenbach, "Two Mass Killings a World Apart Share a Common Theme: 'Ecofascism,'" *Washington Post*, August 18, 2019, https://www.washingtonpost.com/science/two-mass-murders-a-world-apart-share-a-common-theme-ecofascism/2019/08/18/0079a676-bec4-11e9-b873-63ace636af08_story.html.

56. Cited in "Ecofascism: Naomi Klein Warns the Far Right's Embrace of White Supremacy Is Tied to Climate Crisis," *Democracy Now*, September 17, 2019, https://www.democracynow.org/2019/9/17/naomi_klein_eco_fascism.

57. Beth Gardiner, "White Supremacy Goes Green: Why Is the Far Right Suddenly Paying Attention to Climate Change?," *New York Times*, February 28, 2020, https://www.nytimes.com/2020/02/28/opinion/sunday/far-right-climate-change.html. See also John Hultgren, *Border Walls Gone Green: Nature and Anti-Immigrant Politics in America* (Minneapolis: University of Minnesota Press, 2015).

58. Hilary A. Moore, *Burning Earth, Changing Europe: How the Racist Right Exploits the Climate Crisis and What We Can Do About It* (Brussels: Rosa-Luxemburg-Stiftung, 2020), https://www.rosalux.eu/en/article/1588.burning-earth-changing-europe.html.

Chapter 4: Fire as Death

1. Ingrid R.G. Waldron, *There's Something in the Water: Environmental Racism in Indigenous and Black Communities* (Halifax, NS: Fernwood Publishing, 2018).

2. Sarang Narasimhaiah and Mukesh Kulriya, "India's Farmers' Protests: 'This Is History in the Making,'" ROAR, February 5, 2021, https://roarmag.org/essays/indias-farmers-protests/. See also Arundhati Roy, *Capitalism: A Ghost Story* (Chicago: Haymarket Books, 2014).

3. Nick Estes, *Our History Is the Future: Standing Rock Versus the Dakota Access Pipeline, and the Long Tradition of Indigenous Resistance* (London: Verso, 2019); Dina Gilio-Whitaker, *As Long as Grass Grows: The Indigenous Fight for Environmental Justice, from Colonization to Standing Rock* (Boston: Beacon Press, 2019); Julie Sze, *Environmental Justice in a Moment of Danger* (Oakland: University of California Press, 2020).

4. A new battle line in the fight by Indigenous groups and environmentalists against energy companies has emerged around the proposed Line 3 pipeline, a massive project that would pump crude oil from Alberta in Canada through Minnesota to Wisconsin, traversing Native American reservations and fragile ecosystems along the Mississippi River. Activists have demanded that President Biden make good on his promise to stop the expansion of fossil-fuel production and exports of oil and natural gas. Unfortunately, in late June 2021, the Biden administration indicated that it would support the issuing of permits to the Canadian company Enridge Energy and uphold Trump's earlier decision allowing the controversial pipeline to go ahead. See Stephanie Ebbs and MaryAlice Parks, "Line 3 Pipeline Resistance Continues as Activists Ask Biden Admin to Shutdown Project," ABC News, September 23, 2021, https://abcnews.go.com/US/Politics/line-pipeline-resistance-continues-activists-biden-admin-shutdown/story?id=80096664.

5. Winona LaDuke, *The Militarization of Indian Country* (East Lansing: Michigan State University Press, 2013); see also Traci Brynne Voyles, *Wastelanding: Legacies of Uranium Mining in Navajo Country* (Minneapolis: University of Minnesota Press, 2015).

6. Today's militarization of police is a phenomenon that escalated in the United States' reaction to 9/11 and the development of a national antiterrorist response. It was then promoted as a strategy in many countries around the world. It is part of what is often called the military-industrial complex, an alliance between the defense industry, national

176 NOTES TO CHAPTER 4

military, and domestic policing agencies, and is backed politically and financially by national governments. It enables wealthy countries to intervene and defend extractive industries in the global south through the mobilization of land and air force, as well as through the threat of coercive economic sanctions. Related to military interventions overseas, the militarization of domestic police has become evident in many liberal democratic countries around the world, including Australia, Britain, Germany, France, and Canada. Domestic militarization is pushed by weapons producers, who make huge profits on sales. In the United States, for instance, police departments have become a secondary market for weapons sales, and distribution is pushed heavily by defense contractors. The US Department of Defense is authorized to provide surplus military equipment to police departments through the 1033 Program; the departments then have a year to use the equipment or return it. This "use it or lose it" condition encourages police departments to use equipment that is often not appropriate to the issues they are dealing with. Police militarization is also in part driven by public fear. Among white populations, there are heightened concerns about increasing violence, which many associate with racialized minorities and immigrants. Promoting fear was very evident in the political campaigns of Trump, Bolsonaro, and Morrison, who claimed to be protecting citizens from a range of bogeys, including "criminal" Black and Brown youth, "rapist" migrants, and "terrorist" Muslims. Stoking public fear—be it against domestic or foreign "criminals"—is an effective political strategy deployed widely by extreme-right governments. It justifies a range of executive powers and excessive security measures within cities and along borders that shore up ultranationalist ideologies, and is historically linked to fascism and other authoritarian regimes. See Alex S. Vitale, *The End of Policing* (London: Verso, 2018); William I. Robinson, *The Global Police State* (London: Pluto Press, 2020).

7. Vandana Shiva, *Making Peace with the Earth* (London: Pluto Press, 2013), 4.

8. "What Is the Northern Territory Intervention?" See https://www.monash.edu/law/research/centres/castancentre/our-areas-of-work/indigenous/the-northern-territory-intervention/the-northern-territory-intervention-an-evaluation/what-is-the-northern-territory-intervention.

9. Patrick Wolfe, "Settler Colonialism and the Elimination of the Native," *Journal of Genocide Research* 8, no. 4 (2006): 387–409; and Lorenzo Veracini, *Settler Colonialism: A Theoretical Overview* (New York: Palgrave Macmillan, 2010).

10. Bill Gammage, *The Biggest Estate on Earth: How Aborigines Made Australia* (Sydney: Allen & Unwin, 2013).

11. See "A Brief Aboriginal History," Aboriginal Heritage Office, Sydney Australia, accessed January 15, 2021, http://www.aboriginalheri tage.org/history/history/.

12. Similar policies of child removal were implemented by the Canadian government against First Nations and by the United States government against Native Americans. For a moving fictional account of the removal of young Indigenous Australians from their families and the cultural and social impacts, see Kim Scott, *Benang: From the Heart* (Perth: Freemantle Press, 1999).

13. *Western Australia v Ward on Behalf of Miriuwung Gajerrong*, High Court of Australia, 8 August 2002. See "Governments Bow to Mining Companies with Compulsory Acquisition and Biased Legislation," in "Native Title Issues and Problems—Creative Spirits," https:// www.creativespirits.info/aboriginalculture/land/native-title-issues -problems#governments-bow-to-mining-companies-with-compulsory -acquisition-and-biased-legislation.

14. Katie Glaskin, *Crosscurrents: Law and Society in a Native Title Claim to Land and Sea* (Perth: UWAP Scholarly, 2017).

15. Amnesty International, "The Northern Territory and Indigenous Rights," August 20, 2013, https://www.amnesty.org/en /documents/seco1/003/2010/en/.

16. See Shelley Bielefeld, "The Intervention, Stronger Futures, and Racial Discrimination: Placing the Australian Government Under Scrutiny," in *"And There'll Be No Dancing": Perspectives on Policies Impacting Indigenous Australia Since 2007*, ed. Elisabeth Baehr and Barbara Schmidt-Haberkamp, 145–66 (Berlin: Cambridge Scholars Publishing, 2017).

17. Ali Cobby Eckermann, "The Northern Territory Emergency Response: Why Australia Will Not Recover from the Intervention," *Cordite Poetry Review*, February 1, 2015, http://cordite.org.au/essays /the-nt-emergency-response/2/.

18. See Creative Spirits, https://www.creativespirits.info/aborigi nalculture/timeline/searchResults?page=39&q=&category=a. See also Michael Gordon, "Report Claims Intervention Process a Sham," *Sydney Morning Herald*, March 9, 2021, https://www.smh.com.au/politics /federal/report-claims-intervention-process-a-sham-20120308-1un29 .html; and Bielefeld, "The Intervention, Stronger Futures, and Racial Discrimination."

19. Since 1989, the imprisonment rate of Australia's Aboriginal and Torres Strait Islander populations has increased twelve times faster than the rate for non-Aboriginal people. For every 100,000, there are 2,536 Indigenous Australians incarcerated, compared to only 218 non-Indigenous Australians. These figures are even more horrifying given that the Indigenous population constitutes only 3.3 percent of the total Australian population. See "Aboriginal Prison Rates," Creative Spirits, accessed February 12, 2021, https://www.creativespirits.info/aboriginal culture/law/aboriginal-prison-rates.

20. One of the most egregious in this regard is Rio Tinto, the second-largest mining company in the world. In May 2020 it blasted through rock caves in Juukan Gorge in Western Australia containing evidence of human habitation dating back 46,000 years. In this case the unique cultural and archeological site was deliberately destroyed to make way for iron ore extraction. Unbelievably, no laws protect the site from such destruction. However, investors were outraged, and CEO Jean-Sébastien Jacques was forced to resign. Similar destruction of sacred sites by mining companies happens in Canada and the United States. See Kristen A. Carpenter, "Living the Sacred: Indigenous Peoples and Religious Freedom," *Harvard Law Review* 134 (April 2021): 2103–56.

21. MariJo Moore, ed., *Genocide of the Mind: New Native American Writing* (New York: Thunder's Mouth Press / Nation Books, 2003). See also Boaventura de Sousa Santos, *Epistemologies of the South: Justice Against Epistemicide* (London: Routledge, 2014); and Jeffrey S. Bachman, ed., *Cultural Genocide: Law, Politics, and Global Manifestations* (London: Routledge, 2019).

22. Tanya Talaga, *All Our Relations: Finding the Path Forward* (n.p.: House of Anansi Press, 2018); C.E. Eriksen and D.L. Hankins,

"Colonisation and Fire: Gendered Dimensions of Indigenous Fire Knowledge Retention and Revival," in *Routledge Handbook of Gender and Development*, ed. A. Coles, L. Gray, and J. Momsen, 129–37 (New York: Routledge, 2015).

23. For many people, their body and sense of self don't exist autonomously in the world. For instance, Jack Forbes, a leading Indigenous scholar and founder of one of the first Native American Studies programs at UC Davis, writes, "I can lose my hands and still live. I can lose my legs and still live. I can lose my eyes and still live. . . . But if I lose the air I die. If I lose the sun I die. If I lose the earth I die. If I lose the water I die. If I lose the plants and animals I die. All of these things are more a part of me, more essential to my every breath, than is my so-called body. What is my real body?" ("Indigenous Americans: Spirituality and Ecos," *Dædalus* 130, no. 4 [2001]: 283–300). See also Ashlee Cunsolo and Karen Landman, eds., *Mourning Nature: Hope at the Heart of Ecological Loss and Grief* (Montreal, Quebec: McGill-Queen's University Press, 2017).

24. Antony Funnell, "Why the Creeping Militarisation of Our Police Has Experts Worried," ABC News, September 19, 2019, https://www.abc.net.au/news/2019-09-20/creeping-militarisation-of-police-why-experts-are-concerned/11517266#:~:text=Australian%20police%20are%20increasingly%20being,convergence%22%20as%20slow%20and%20worrying.

25. Australia's refugee laws, detention centers, and offshoring of refugees on distant islands have been praised by Trump and other antidemocratic leaders as a model system for dealing with the pressures of mass migration. But Australia has been widely condemned by the United Nations and other international watchdogs for its violation of asylum seekers' basic human rights that typically leads to indefinite detention without legal representation and appropriate processing of applications. Some refugees have been held in detention for decades. Luke Henriques-Gomes, "Donald Trump Says 'Much Can Be Learned' from Australia's Hardline Asylum Seeker Policies," *Guardian*, June 26, 2019, https://www.theguardian.com/us-news/2019/jun/27/donald-trump-says-much-can-be-learned-from-australias-hardline-asylum-seeker-policies; Ben Doherty, "UN Body Condemns Australia

for Illegal Detention of Asylum Seekers and Refugees," *Guardian*, July 7, 2018, https://www.theguardian.com/world/2018/jul/08/un -body-condemns-australia-for-illegal-detention-of-asylum-seekers-and -refugees.

26. Business and Human Rights Resource Center, "Paraguay: UN Human Rights Committee Rules State Failure to Prevent Toxic Agribusiness Pollution Violates Indigenous Rights," https://www.business -humanrights.org/en/.

27. See David Price, *Before the Bulldozer: The Nambiquara Indians and the World Bank* (Washington, DC: Seven Locks Press, 1989); and Chris Feliciano Arnold, *The Third Bank of the River: Power and Survival in the Twenty-First Century Amazon* (New York: Picador, 2018).

28. Fiona Watson, "Bolsonaro's Election Is Catastrophic News for Brazil's Indigenous Tribes," *Guardian*, October 31, 2018, https://www .theguardian.com/commentisfree/2018/oct/31/jair-bolsonaro-brazil -indigenous-tribes-mining-logging; M.B. Sattar, *Greed and Politics Is Burning Down the Lungs of the Earth: The Curious Case of Wildfires in Amazon Rainforest* (San Bernardino, CA: n.p., 2020), 44.

29. Cited in Dom Phillips, "'Project of Death': Alarm at Bolsonaro's Plan for Amazon-Spanning Bridge," *Guardian*, March 10, 2020, https://www.theguardian.com/global-development/2020/mar/10 /brazil-amazon-bridge-project-bolsonaro.

30. Cited in Ernesto Londoño and Leticia Casado, "Under Pressure, Brazil's President Moves to Fight Deforestation," *New York Times*, August 2, 2020, https://www.nytimes.com/2020/08/01/world/amer icas/Brazil-amazon-deforestation-bolsonaro.html.

31. For an English translation of the original letter written in Portuguese to the global community, see Rainforest Rescue, https://www .rainforest-rescue.org/petitions/1071/a-cry-for-help-from-the-amazon -rainforest.

32. Amy Goodman, "Brazilian Indigenous Leader Davi Kopenawa: Bolsonaro Is Killing My People and Destroying the Amazon," *Democracy Now*, December 4, 2019, https://www.democracynow.org/2019 /12/4/yanomami_indigenous_leader_protecting_amazon.

33. "Report: Violência contra os povos indígenas no Brasil, dados de 2018" (Report: Violence Against Indigenous Peoples in Brazil,

2018 Data), Conselho Indigenista Missionário (Cimi), https://cimi.org
.br/wp-content/uploads/2019/09/relatorio-violencia-contra-os-povos
-indigenas-brasil-2018.pdf.

34. Cited in Leonardo Benassatto and Ueslei Marcelino, "Fighting
Fire with Fire, Amazon 'Forest Guardians' Stalk Illegal Loggers," Reuters, September 20, 2019, https://www.reuters.com/article/uk-brazil
-environment-forest-guardians-w-idUKKBN1W51DO.

35. Tricontinental: Institute for Social Research, *Popular Agrarian Reform and the Struggle for Land in Brazil*, dossier no. 27, April
2020, https://www.thetricontinental.org/wp-content/uploads/2020
/04/20200328_Dossier-27_EN_Web.pdf.

36. Tom Philpott, *Perilous Bounty: The Looming Collapse of American Farming and How We Can Prevent It* (New York: Bloomsbury,
2020).

37. California Research Bureau, "Farmworkers in California: A
Brief Introduction," S-13-017, October 2013, https://latinocaucus
.legislature.ca.gov/sites/latinocaucus.legislature.ca.gov/files/CRB
%20Report%20on%20Farmworkers%20in%20CA%20S-13-017.pdf.

38. Christopher Giles, "Trump's Wall: How Much Has Been Built
During His Term?," BBC News, January 12, 2021, https://www.bbc
.com/news/world-us-canada-46748492; Public Policy Institute of California, "Fact Sheet: Immigrants in California," March 2021, https://
www.ppic.org/publication/immigrants-in-california/.

39. Throughout 2020, ICE granted a temporary order for field-workers deemed "essential workers" in the context of the COVID-19
pandemic, which allowed them to leave their homes despite lockdown
orders. But essential worker status meant that undocumented immigrants could still be detained and deported at any moment. This disingenuous move by the Trump administration helped maintain food supply for the nation without providing those at risk with access to health
services or formal avenues to citizenship.

40. Seth Holmes, *Fresh Fruit, Broken Bodies: Migrant Farmworkers in the United States* (Berkeley: University of California Press, 2013).

41. Lori A. Flores, *Grounds for Dreaming: Mexican Americans,
Mexican Immigrants, and the California Farmworker Movement*, Lamar Series in Western History (New Haven, CT: Yale University Press,

2016). See also Center for Farmworker Families: Education, Advocacy, and Support (Felton, CA), accessed January 19, 2021, https://farm workerfamily.org/information.

42. Cited in Vivian Ho, "'An Impossible Choice': Farmworkers Pick a Paycheck over Health Despite Smoke-Filled Air," *Guardian*, August 23, 2020, https://www.theguardian.com/us-news/2020/aug/22/california-farmworkers-wildfires-air-quality-coronavirus.

43. Cited in Ho, "Impossible Choice."

44. Somini Sengupta, "Heat, Smoke, and COVID Are Battering the Workers Who Feed America," *New York Times*, August 12, 2020, https://www.nytimes.com/2020/08/25/climate/california-farm-work ers-climate-change.html.

45. Michael Méndez et al., "The (In)Visible Victims of Disaster: Understanding the Vulnerability of Undocumented Latino/a and Indigenous Immigrants," *Geoforum* 116 (2020): 50–62.

46. Another dimension is the cheap labor provided by prisoners that crowd California's prison-industrial complex and who are often conscripted into fighting wildfires. Black and Brown youth are disproportionately represented in the state's jails and provide another silenced and exploited labor force benefiting capitalist enterprises. See Ruth Wilson Gilmore, *Golden Gulag: Prisons, Surplus, Crisis, and Opposition in Globalizing California* (Berkeley: University of California Press, 2007).

47. Almond, R.E.A., M. Grooten, and T. Petersen, eds., *Living Planet Report 2020: Bending the Curve of Biodiversity Loss* (Gland, Switzerland: World Wildlife Foundation, 2020), https://f.hubspotusercon tent20.net/hubfs/4783129/LPR/PDFs/ENGLISH-FULL.pdf.

Chapter 5: Fire as Disruption

1. Thomas Fuller, "Wildfires Blot Out Sun in the Bay Area," *New York Times*, September 9, 2020, https://www.nytimes.com/2020/09/09/us/pictures-photos-california-fires.html.

2. Cited in Michael Slezak, "Three Billion Animals Killed or Displaced in Black Summer Bushfires, Study Estimates," ABC News, July 27, 2020, https://www.abc.net.au/news/2020-07-28/3-billion-animals-killed-displaced-in-fires-wwf-study/12497976.

3. Alicia Garza, one of the creators of the #BlackLivesMatter hashtag, explained how Black lives mattering is a precondition for all

lives mattering: "Black Lives Matter doesn't mean your life isn't important—it means that Black lives, which are seen as without value within
White supremacy, are important to your liberation. Given the disproportionate impact state violence has on Black lives, we understand that
when Black people in this country get free, the benefits will be wide-
reaching and transformative for society as a whole. When we are able
to end the hyper-criminalization and sexualization of Black people and
end the poverty, control, and surveillance of Black people, every single
person in this world has a better shot at getting and staying free. When
Black people get free, everybody gets free." See Garza, "A Herstory
of the #BlackLivesMatter Movement," Feminist Wire, October 2014,
http://www.thefeministwire.com/2014/10/blacklivesmatter-2/.

4. See the 2019 Global Assessment Report on Biodiversity and Ecosystem Services (https://en.wikipedia.org/wiki/Global_Assessment
_Report_on_Biodiversity_and_Ecosystem_Services) published by the
United Nations Intergovernmental Science-Policy Platform on Biodiversity and Ecosystem Services (https://en.wikipedia.org/wiki/Inter
governmental_Science-Policy_Platform_on_Biodiversity_and_Eco
system_Services).

5. Matthew Taylor, Jonathan Watts, and John Bartlett, "Climate
Crisis: Six Million People Join Latest Wave of Global Protests," *Guardian*, September 27, 2019, https://www.theguardian.com/environment
/2019/sep/27/climate-crisis-6-million-people-join-latest-wave-of
-worldwide-protests.

6. World Economic Forum, "Our Mission," https://www.weforum
.org/about/world-economic-forum.

7. "Davos 2020: Prince Charles Meets Greta Thunberg," BBC
News, September 22, 2020, https://www.bbc.com/news/uk-51211853.

8. "Greta Thunberg's Remarks at the Davos Economic Forum,"
New York Times, January 21, 2021, https://www.nytimes.com/2020
/01/21/climate/greta-thunberg-davos-transcript.html; Kate Whiting,
"Prince Charles: Ten Actions We Must Take to Drive the Green Recovery," World Economic Forum, November 10, 2020, https://www
.weforum.org/agenda/2020/11/prince-charles-the-10-actions-we-must
-take-to-tackle-climate-change/.

9. The idea of a separation between wildlife and people was the
underlying assumption of the conservation movement established in

Britain and Germany in the mid-nineteenth century before taking off in the United States, inspired by the writings of Henry David Thoreau and led by future president Theodore Roosevelt. In 1872 Yellowstone National Park was established, the first of many national parks set up to preserve unique pristine natural habitats for the pleasure and leisure of future (white) generations. Today, critical conservation scholars are challenging the human/nature binary and calling for new ways of thinking about the interconnected life of humans and nonhumans. For a fascinating discussion of the unfolding global conservation debates in the context of neoliberalism, see Bram Büscher and Robert Fletcher, *The Conservation Revolution: Radical Ideas for Saving Nature Beyond the Anthropocene* (London: Verso, 2020); see also Taylor, *The Rise of the American Conservation Movement*; Dowie, *Conservation Refugees*; and Curnow and Helferty, "Contradictions of Solidarity."

10. Cited in Isla Binnie, "Activist Thunberg Turns Spotlight on Indigenous Struggle at Climate Summit," Reuters, December 9, 2019, https://www.reuters.com/article/us-climate-change-accord-greta/activist-thunberg-turns-spotlight-on-indigenous-struggle-at-climate-summit-idUSKBN1YD1J5.

11. Global Witness Report, *Defending Tomorrow*, July 29, 2020, https://www.globalwitness.org/en/campaigns/environmental-activists/defending-tomorrow/.

12. Mélissa Godin, "Record Number of Environmental Activists Killed in 2019," *Time*, July 29, 2020, https://time.com/5873137/record-number-killing-environmental-activists-2019/.

Appendix

1. Nadja Popovich, Livia Albeck-Ripka, and Kendra Pierre-Louis, "The Trump Administration Rolled Back More Than 100 Environmental Rules. Here's the Full List," *New York Times*, updated January 20, 2021, https://www.nytimes.com/interactive/2020/climate/trump-environment-rollbacks-list.html#:~:text=All%20told%2C%20the%20Trump%20administration's,to%20energy%20and%20legal%20analysts.

BIBLIOGRAPHY

Almond, R.E.A., M. Grooten, and T. Petersen, eds. *Living Planet Report 2020: Bending the Curve of Biodiversity Loss*. Gland, Switzerland: World Wildlife Foundation, 2020. https://f.hubspotusercontent20 .net/hubfs/4783129/LPR/PDFs/ENGLISH-FULL.pdf.

Amnesty International. "The Northern Territory and Indigenous Rights." August 20, 2013. https://www.amnesty.org/en/documents /sec01/003/2010/en/.

Anderson, Perry. *Brazil Apart: 1964–2019*. London: Verso, 2019.

Applebaum, Anne. *Twilight of Democracy: The Seductive Lure of Authoritarianism*. New York: Doubleday, 2020.

Arnold, Chris Feliciano. *The Third Bank of the River: Power and Survival in the Twenty-First Century Amazon*. New York: Picador, 2018.

Ayers, Alison J., and Alfredo Saad-Filho. "Democracy Against Neoliberalism: Paradoxes, Limitations, Transcendence." *Critical Sociology* 41, nos. 4–5 (2015): 597–618.

Bachman, Jeffrey S., ed. *Cultural Genocide: Law, Politics, and Global Manifestations*. London: Routledge, 2019.

Baer, Hans A. *Global Capitalism and Climate Change: The Need for an Alternative World System*. Lanham, MD: AltaMira Press, 2012.

Baldwin, James. *The Fire Next Time*. New York: Vintage, 1992.

Beck, Ulrich. *The Metamorphosis of the World: How Climate Change Is Transforming Our Concept of the World.* Cambridge, UK: Polity Press, 2016.

Bello, Walden. *Counterrevolution: The Global Rise of the Far Right.* Black Point, NS: Fernwood Publishing, 2020.

Berberoglu, Berch. *The Global Rise of Authoritarianism in the 21st Century: Crisis of Neoliberal Globalization and the Nationalist Response.* London: Routledge, 2020.

Biehl, Janet, and Peter Staudenmaier. *Ecofascism Revisited: Lessons from the German Experience.* 2nd ed. Porsgrunn, Norway: New Compass Press, 2011.

Bielefeld, Shelley. "The Intervention, Stronger Futures, and Racial Discrimination: Placing the Australian Government Under Scrutiny." In *"And There'll Be No Dancing": Perspectives on Policies Impacting Indigenous Australia Since 2007,* edited by Elisabeth Baehr and Barbara Schmidt-Haberkamp, 145–66. Berlin: Cambridge Scholars Publishing, 2017.

Black, Megan. *The Global Interior: Mineral Frontiers and American Power.* Cambridge, MA: Harvard University Press, 2018.

Boon, Sonja. *What the Oceans Remember: Searching for Belonging and Home.* Waterloo, ON: Wilfrid Laurier University Press, 2019.

Borras, Saturnino M., Philip Seufert, Stephan Backes, Daniel Fyfe, Laura Michele, and Elyse Mills. *Land Grabbing and Human Rights: The Involvement of European Corporate and Financial Entities in Land Grabbing Outside the European Union.* European Parliament's Subcommittee on Human Rights, 2016. https://www.europarl.europa.eu/RegData/etudes/STUD/2016/578007/EXPO_STU(2016)578007_EN.pdf.

Boucher, Doug, Sarah Roquemore, and Estrellita Fitzhugh. "Brazil's Success in Reducing Deforestation." *Tropical Conservation Science* 6, no. 3 (2013): 426–45.

Bradshaw, Corey J.A., and Paul R. Ehrlich. *Killing the Koala and Poisoning the Prairie: Australia, America, and the Environment.* Chicago: University of Chicago Press, 2015.

Braverman, Irus, and Elizabeth R. Johnson, eds. *Blue Legalities: The Life and Laws of the Sea.* Durham, NC: Duke University Press, 2020.

Brown, Faye. "Smoke from Australia's Wildfires Has Travelled over 9,000 Miles to Brazil." *Metro*, January 10, 2020. https://metro.co .uk/2020/01/10/smoke-australias-wildfires-travelled-9000-miles -brazil-12036242/.

Brown, Wendy. *In the Ruins of Neoliberalism: The Rise of Antidemocratic Politics in the West.* New York: Columbia University Press, 2019.

———. *Undoing the Demos: Neoliberalism's Stealth Revolution.* London: Zone Books, 2015.

Browning, Christopher R. "The Suffocation of Democracy." *New York Review of Books*, October 25, 2018. https://www.nybooks.com/arti cles/2018/10/25/suffocation-of-democracy/.

Bruff, Ian. "The Rise of Authoritarian Neoliberalism." *Rethinking Marxism* 26, no. 1 (2014): 113–29.

Büscher, Bram, and Robert Fletcher. *The Conservation Revolution: Radical Ideas for Saving Nature Beyond the Anthropocene.* London: Verso, 2020.

Carpenter, Kristen A. "Living the Sacred: Indigenous Peoples and Religious Freedom." *Harvard Law Review* 134 (April 2021): 2103–56.

Carson, Rachel. *Silent Spring.* 40th anniversary ed. Boston: Houghton Mifflin, 2002.

Chomsky, Noam. *Profit over People: Neoliberalism and Global Order.* New York: Seven Stories Press, 1999.

Coleman, Charles M. *P.G. and E. of California: The Centennial Story of Pacific Gas and Electric Company, 1852–1952.* New York: McGraw-Hill, 1952.

Connolly, William E. *Aspirational Fascism: The Struggle for Multifaceted Democracy Under Trumpism.* Minneapolis: University of Minnesota Press, 2017.

Corrales, Javier. "The Authoritarian Resurgence: Autocratic Legalism in Venezuela." *Journal of Democracy* 26, no. 2 (2015): 37–51.

Cronon, William. *Changes in the Land: Indians, Colonists, and the Ecology of New England.* Rev. ed. New York: Hill and Wang, 2003.

Crutzen, Paul J. "Geology of Mankind: The Anthropocene." *Nature* 415 (2002): 23.

Cunsolo, Ashlee, and Karen Landman, eds. *Mourning Nature: Hope at the Heart of Ecological Loss and Grief.* Montreal, Quebec: McGill-Queen's University Press, 2017.

Curnow, Joe, and Anjali Helferty. "Contradictions of Solidarity: Whiteness, Settler Coloniality, and the Mainstream Environmental Movement." *Environment and Society* 9, no. 1 (2018): 145–63.

Darian-Smith, Eve. "Dying for the Economy: Disposable People and Economies of Death in the Global North." *State Crime Journal* 10, no. 1 (2021): 61–79.

———. "Global Law Firms in Real-World Contexts: Practical Limitations and Ethical Implications." *Beijing Law Review* 6 (2015): 92–101.

———. "Who Owns the World? Landscapes of Sovereignty, Property, Dispossession." *Journal of the Oxford Centre for Socio-Legal Studies* 1, no. 2 (2016). https://www.law.ox.ac.uk/content/annual-socio -legal-lecture-who-owns-world-landscapes-sovereignty-property -dispossession.

Darian-Smith, Eve, and Philip C. McCarty. *The Global Turn: Theories, Research Designs, and Methods for Global Studies.* Oakland: University of California Press, 2017.

Davis, Mike. *Ecology of Fear: Los Angeles and the Imagination of Disaster.* New York: Vintage, 1999.

Demissie, Fassil, ed. *Land Grabbing in Africa: The Race for Africa's Rich Farmland.* New York: Routledge, 2017.

Dowie, Mark. *Conservation Refugees: The Hundred-Year Conflict Between Global Conservation and Native Peoples.* Cambridge, MA: MIT Press, 2011.

Duggan, Lisa. *The Twilight of Equality? Neoliberalism, Cultural Politics, and the Attack on Democracy.* Boston: Beacon Press, 2003.

Dunlap, Riley E., and Peter J. Jacques. "Climate Change Denial Books and Conservative Think Tanks: Exploring the Connection." *American Behavioral Scientist* 57, no. 6 (2013): 699–731.

Eriksen, C.E., and D.L. Hankins. "Colonisation and Fire: Gendered Dimensions of Indigenous Fire Knowledge Retention and Revival." In *The Routledge Handbook of Gender and Development*, edited by A. Coles, L. Gray, and J. Momsen, 129–37. New York: Routledge, 2015.

Estes, Nick. *Our History Is the Future: Standing Rock Versus the Dakota Access Pipeline, and the Long Tradition of Indigenous Resistance.* London: Verso, 2019.

Ferguson, Gary. *Land on Fire: The New Reality of Wildfire in the West.* Portland, OR: Timber Press, 2017.

Flores, Lori A. *Grounds for Dreaming: Mexican Americans, Mexican Immigrants, and the California Farmworker Movement.* Lamar Series in Western History. New Haven, CT: Yale University Press, 2016.

Forbes, Jack D. "Indigenous Americans: Spirituality and Ecos." *Dædalus* 130, no. 4 (2001): 283–300.

Forchtner, Bernhard, ed. *The Far Right and the Environment: Politics, Discourse, and Communication.* New York: Routledge, 2020.

Foster, John Bellamy. *The Return of Nature: Socialism and Ecology.* New York: Monthly Review Press, 2020.

Freese, Barbara. *Industrial-Strength Denial: Eight Stories of Corporations Defending the Indefensible, from the Slave Trade to Climate Change.* Oakland: University of California Press, 2020.

Gammage, Bill. *The Biggest Estate on Earth: How Aborigines Made Australia.* Sydney: Allen & Unwin, 2013.

Gee, Alistair, and Dani Anguiano. *Fire in Paradise: An American Tragedy.* New York: W.W. Norton, 2020.

Gershon, Livia. "The Global Suppression of Indigenous Fire Management." JSTOR Daily, October 12, 2020. https://daily.jstor.org/the-global-suppression-of-indigenous-fire-management/.

Gilio-Whitaker, Dina. *As Long as Grass Grows: The Indigenous Fight for Environmental Justice, from Colonization to Standing Rock.* Boston: Beacon Press, 2019.

Gilmore, Ruth Wilson. *Golden Gulag: Prisons, Surplus, Crisis, and Opposition in Globalizing California.* Berkeley: University of California Press, 2007.

———. "Race and Globalization." In *Geographies of Global Change*, edited by Peter J. Taylor, R.L. Johnstone, and Michael J. Watts, 261–74. 2nd ed. Oxford, UK: Blackwell, 2002.

Giroux, Henry A. *The Terror of Neoliberalism: Authoritarianism and the Eclipse of Democracy.* Boulder, CO: Paradigm Publishers, 2004.

Glaskin, Katie. *Crosscurrents: Law and Society in a Native Title Claim to Land and Sea.* Perth: UWAP Scholarly, 2017.

Goodman, Amy. "Brazilian Indigenous Leader Davi Kopenawa: Bolsonaro Is Killing My People and Destroying the Amazon." *Democracy Now.* December 4, 2019. https://www.democracynow.org/2019/12/4/yanomami_indigenous_leader_protecting_amazon.

GRAIN. "Seized: The 2008 Landgrab for Food and Financial Security." October 24, 2008. Accessed February 2, 2021. https://www.grain.org/article/entries/93-seized-the-2008-landgrab-for-food-and-financial-security.pd.

Gramling, Carolyn. "Four Ways to Put the 100-Degree Arctic Heat Record in Context." *Science News,* July 1, 2020. https://www.sciencenews.org/article/climate-new-high-temperature-heat-record-arctic-siberia-context.

Halliday, Terence C., and Gregory Shaffer, eds. *Transnational Legal Orders.* Cambridge: Cambridge University Press, 2016.

Hamilton, Clive. *Requiem for a Species: Why We Resist the Truth About Climate Change.* New York: Routledge, 2010.

———. *Silencing Dissent: How the Australian Government Is Controlling Public Opinion and Stifling Debate.* Melbourne: Allen & Unwin, 2007.

Haraway, Donna J. *Staying with the Trouble: Making Kin in the Chthulucene.* Durham, NC: Duke University Press, 2016.

———. *When Species Meet.* Minneapolis: University of Minnesota Press, 2008.

Harvey, David. *A Brief History of Neoliberalism.* Oxford: Oxford University Press, 2007.

Hau'ofa, Epeli. "The Ocean in Us." *The Contemporary Pacific* 10, no. 2 (1998): 392–410.

Henson, Bob. "Silent Calamity: The Health Impacts of Wildfire Smoke." *Yale Climate Connections,* May 21, 2021. https://yaleclimateconnections.org/2021/05/silent-calamity-the-health-impacts-of-wildfire-smoke/.

Hermansson, Patrik, David Lawrence, Joe Mulhall, and Simon Murdoch. *The International Alt-Right: Fascism for the 21st Century?* New York: Routledge, 2020.

Hickel, Jason. *The Divide: Global Inequality from Conquest to Free Markets.* New York: W.W. Norton, 2018.

———. *Less Is More: How Degrowth Will Save the World*. London: William Heinemann, 2020.

Hinton, Elizabeth. *America on Fire: The Untold History of Police Violence and Black Rebellion Since the 1960s*. New York: Liveright, 2021.

Hochschild, Arlie Russell. *Strangers in Their Own Land: Anger and Mourning on the American Right*. New York: New Press, 2018.

Hoelle, Jeffrey. *Rainforest Cowboys: The Rise of Ranching and Cattle Culture in Western Amazonia*. Austin: University of Texas Press, 2015.

Hoffman, Andrew. *How Culture Shapes the Climate Change Debate*. Stanford, CA: Stanford University Press, 2015.

Holmes, Seth. *Fresh Fruit, Broken Bodies: Migrant Farmworkers in the United States*. Berkeley: University of California Press, 2013.

Horesh, Theo. *The Fascism This Time: And the Global Future of Democracy*. Boulder, CO: Cosmopolis Press, 2020.

Horne, Gerald. *The Dawning of the Apocalypse: The Roots of Slavery, White Supremacy, Settler Colonialism, and Capitalism in the Long Sixteenth Century*. New York: Monthly Review Press, 2020.

Huber, Robert A. "The Role of Populist Attitudes in Explaining Climate Change Skepticism and Support for Environmental Protection." *Environmental Politics* 29, no. 6 (2020): 959–82.

Hultgren, John. *Border Walls Gone Green: Nature and Anti-Immigrant Politics in America*. Minneapolis: University of Minnesota Press, 2015.

Ingersoll, Karin Amimoto. *Waves of Knowing: A Seascape Epistemology*. Durham, NC: Duke University Press, 2016.

Jackson, Jack. *Law Without Future: Anti-Constitutional Politics and the American Right*. Philadelphia: University of Pennsylvania Press, 2019.

Judis, John B. *The Nationalist Revival: Trade, Immigration, and the Revolt Against Globalization*. New York: Columbia Global Reports, 2018.

Juego, Bonn. "Authoritarian Neoliberalism: Its Ideological Antecedents and Policy Manifestations from Carl Schmitt's Political Economy of Governance." *Administrative Culture* 19, no. 1 (2018): 105–36.

Kimmerer, Robin Wall. *Braiding Sweetgrass: Indigenous Wisdom, Scientific Knowledge, and the Teachings of Plants.* Minneapolis: Milkweed Editions, 2013.

Klein, Naomi. *On Fire: The (Burning) Case for a Green New Deal.* New York: Allen Lane, 2019.

———. *This Changes Everything: Capitalism vs. the Climate.* New York: Simon & Schuster Paperbacks, 2014.

Knox, Malcolm. *Boom: The Underground History of Australia, from Gold Rush to GFC.* Melbourne: Viking, 2013.

Kodas, Michael. *Megafire: The Race to Extinguish a Deadly Epidemic of Flame.* Boston: Houghton Mifflin Harcourt, 2017.

Kojola, Erik. "Bringing Back the Mines and a Way of Life: Populism and the Politics of Extraction." *Annals of the American Association of Geographers* 109, no. 2 (2019): 371–81.

Kolbert, Elizabeth. *The Sixth Extinction: An Unnatural History.* New York: Picador, 2015.

———. *Under a White Sky: The Nature of the Future.* New York: Crown, 2021.

Krugman, Paul. *Arguing with Zombies: Economics, Politics, and the Fight for a Better Future.* New York: W.W. Norton, 2020.

LaDuke, Winona. *The Militarization of Indian Country.* East Lansing: Michigan State University Press, 2013.

Larsen, Soren C., and Jay T. Johnson. *Being Together in Place: Indigenous Coexistence in a More Than Human World.* Minneapolis: University of Minnesota Press, 2017.

Latour, Bruno. *Down to Earth: Politics in the New Climatic Regime.* Cambridge, UK: Polity, 2017.

Layton-Bennett, Anne. "Dear Mr. Morrison." In *From the Ashes: A Poetry Collection in Support of the 2019–2020 Australian Bushfire Relief Effort,* edited by C.S. Huges, 61. Korumburra: Maximum Felix Media, 2020.

Lemon, Don. *This Is the Fire: What I Say to My Friends About Racism.* New York: Little, Brown, 2021.

Lerner, Steve. *Sacrifice Zones: The Front Lines of Toxic Chemical Exposure in the United States.* Cambridge, MA: MIT Press, 2010.

Levitsky, Steven, and Daniel Ziblatt. *How Democracies Die.* New York: Crown, 2018.

Lewis, Simon L., and Mark A. Maslin. "Defining the Anthropocene." *Nature* 519 (2015): 171–80.

Liberti, Stefano. *Land Grabbing: Journeys in the New Colonialism.* Translated by Enda Flannelly. London: Verso, 2013.

Likhani, Nina. *Who Killed Berta Caceres? Dams, Death Squads, and an Indigenous Defender's Battle for the Planet.* London: Verso, 2020.

Linz, Juan. *Totalitarian and Authoritarian Regimes.* Boulder, CO: Lynne Rienner Publishers, 2000.

Lockwood, Matthew. "Right-Wing Populism and the Climate Change Agenda: Exploring Linkages." *Environmental Politics* 27, no. 4 (2018): 712–32.

MacLean, Nancy. *Democracy in Chains: The Radical Right's Stealth Plan for America.* New York: Penguin Books, 2018.

Mann, Michael. *The New Climate War: The Fight to Take Back Our Planet.* New York: Public Affairs, 2021.

Mann, Michael, and Tom Toles. *The Madhouse Effect: How Climate Change Denial Is Threatening Our Planet, Destroying Our Politics, and Driving Us Crazy.* New York: Columbia University Press, 2018.

Mawani, Renisa. *Across Oceans of Law: The Komagata Maru and Jurisdiction in the Time of Empire.* Durham, NC: Duke University Press, 2018.

Mayer, Jane. *Dark Money: The Hidden History of the Billionaires Behind the Rise of the Radical Right.* New York: Penguin, 2017.

Maza, Christina. "Far-Right Climate Denial Is Growing in Europe." *New Republic*, November 11, 2019.

McLauchlan, Kendra K., Philip E. Higuera, Jessica Miesel, Brendan M. Rogers, Jennifer Schweitzer, Jacquelyn K. Shuman, Alan J. Tepley, et al. "Fire as a Fundamental Ecological Process: Research Advances and Frontiers." *Journal of Ecology* 108 (2020): 2047–69.

McLean, Ian W. *Why Australia Prospered: The Shifting Sources of Economic Growth.* Princeton, NJ: Princeton University Press, 2016.

Meadows, Donella H., Dennis Meadows, Jørgen Randers, and William W. Behrens III. *The Limits to Growth: A Report for the Club of Rome's Project on the Predicament of Mankind.* New York: Universe Books, 1972.

Méndez, Michael, Genevieve Flores-Haro, and Lucas Zucker. "The (In)Visible Victims of Disaster: Understanding the Vulnerability of

Undocumented Latino/a and Indigenous Immigrants." *Geoforum* 116 (2020): 50–62.

Merrick, Helen. "Naturecultures and Feminist Materialism." In *The Routledge Handbook of Gender and Environment*, edited by Sherilyn MacGregor, 101–14. New York: Routledge, 2017.

Mirowski, Philip, and Dieter Plehwe. *The Road from Mount Pèlerin: The Making of the Neoliberal Thought Collective*. Cambridge, MA: Harvard University Press, 2009.

Mitchell, Timothy. *Carbon Democracy: Political Power in the Age of Oil*. London: Verso, 2013.

———. *Rule of Experts: Egypt, Techno-Politics, Modernity*. Berkeley: University of California Press, 2002.

Mondon, Aurélien. *The Mainstreaming of the Extreme Right in France and Australia: A Populist Hegemony*. New York: Routledge, 2016.

Moore, Hilary A. *Burning Earth, Changing Europe: How the Racist Right Exploits the Climate Crisis and What We Can Do About It*. Brussels: Rosa-Luxemburg-Stiftung, 2020. https://www.rosalux.eu/en/article/1588.burning-earth-changing-europe.html.

Moore, Jason W. *Capitalism in the Web of Life: Ecology and the Accumulation of Capital*. London: Verso, 2015.

Moore, MariJo, ed. *Genocide of the Mind: New Native American Writing*. New York: Thunder's Mouth Press / Nation Books, 2003.

Morrison, Kathleen D. "Provincializing the Anthropocene." *Seminar: The Monthly Symposium* 673 (September 2015, Nature and History): 75–80.

Myers, Niels M., and Jorgen S. Nørgård. "Policy Means for Sustainable Energy Scenarios." In "Collection of Extended Abstracts," *Proceedings of the International Conference on Energy, Environment and Health—Optimisation of Future Energy Systems*, May 31–June 2, 2010, Carlsberg Academy, Copenhagen, Denmark, 133–37. https://web.archive.org/web/20161009062231/http://www2.dmu.dk/1_om_dmu/2_afdelinger/3_atmi/ceeh/Collection%20of%20Extended%20Abstracts.pdf#page=134.

Nixon, Rob. *Slow Violence and the Environmentalism of the Poor*. Cambridge, MA: Harvard University Press, 2011.

Norris, Pippa, and Ronald Inglehart. *Cultural Backlash: Trump, Brexit, and the Rise of Authoritarian-Populism*. New York: Cambridge University Press, 2019.

Oliveira, Gustavo de L. T., and Mindi Schneider. "The Politics of Flexing Soybeans: China, Brazil and Global Agroindustrial Restructuring." *Journal of Peasant Studies* 43, no. 1 (2016): 167–94.

Oreskes, Naomi, and Erick M. Conway. *Merchants of Doubt: How a Handful of Scientists Obscured the Truth on Issues from Tobacco Smoke to Global Warming.* London: Bloomsbury Press, 2010.

Parenti, Christian. "'The Limits to Growth': A Book That Launched a Movement." *Nation*, December 5, 2012. https://www.thenation.com/article/archive/limits-growth-book-launched-movement/.

Park, Sora, Caroline Fisher, Jee Young Lee, Kieran McGuinness, Yoonmo Sang, Mathieu O'Neil, Michael Jensen, et al. "Digital News Report: Australia 2020." News and Media Research Centre, University of Canberra. https://apo.org.au/node/305057.

Peck, Jamie. *Constructions of Neoliberal Reason.* Oxford: Oxford University Press, 2010.

Philpott, Tom. *Perilous Bounty: The Looming Collapse of American Farming and How We Can Prevent It.* New York: Bloomsbury, 2020.

Prashad, Vijay, ed. *Strongmen: Trump-Modi-Erdogan-Duterte-Putin.* New Delhi: LeftWord, 2018.

Price, David. *Before the Bulldozer: The Nambiquara Indians and the World Bank.* Washington, DC: Seven Locks Press, 1989.

Pulido, Laura. "Rethinking Environmental Racism: White Privilege and Urban Development in Southern California." *Annals of the Association of American Geographers* 90, no. 1 (2000): 12–40.

Pulido, Laura, Tianna Bruno, Cristina Falver-Serna, and Cassandra Galentine. "Environmental Deregulation, Spectacular Racism, and White Nationalism in the Trump Era." *Annals of the American Association of Geographers* 109, no. 2 (2019): 520–32.

Pyne, Stephen J. *The Pyrocene: How We Created an Age of Fire, and What Happens Next.* Oakland: University of California Press, 2021.

Repucci, Sarah. *Freedom in the World 2020: A Leaderless Struggle for Democracy.* Freedom House Report, 2020. Accessed February 16, 2021. https://freedomhouse.org/sites/default/files/2020-02/FIW_2020_REPORT_BOOKLET_Final.pdf.

Riley, Sharon J. "Alberta Suspends at Least 19 Monitoring Requirements in Oilsands, Citing Coronavirus Concerns." *Narwhal*, May 5, 2020.

https://thenarwhal.ca/alberta-suspends-19-oilsands-environmental
-monitoring-requirements-coronavirus-concerns/.

Robinson, Cedric. *Black Marxism: The Making of the Black Radical Tradition.* London: Zed Press, 1983.

Robinson, William I. *The Global Police State.* London: Pluto Press, 2020.

Rome, Adam. *The Genius of Earth Day: How a 1970 Teach-In Unexpectedly Made the First Green Generation.* New York: Hill and Wang, 2014.

Rosane, Olivia. "German Far Right Party Attacks 16-Year-Old Greta Thunberg Ahead of EU Elections." EcoWatch, May 15, 2019. https://www.ecowatch.com/german-far-right-party-greta-thunberg-2637190985.html.

Ross, Alexander Reid, ed. *Grabbing Back: Essays Against the Global Land Grab.* Oakland, CA: AK Press, 2014.

Rothschild, Mike. *The Storm Is upon Us: How QAnon Became a Movement, Cult, and Conspiracy Theory of Everything.* Brooklyn, NY: Melville House, 2021.

Roy, Arundhati. *Capitalism: A Ghost Story.* Chicago: Haymarket Books, 2014.

Rydgren, Jens, ed. *The Oxford Handbook of the Radical Right.* Oxford: Oxford University Press, 2018.

Saad-Filho, Alfredo, and Lecio Morais. *Brazil: Neoliberalism Versus Democracy.* London: Pluto Press, 2018.

Saito, Kohei. *Karl Marx's Ecosocialism: Capitalism, Nature, and the Unfinished Critique of Political Economy.* New York: Monthly Review Press, 2017.

Santos, Boaventura de Sousa. *The End of the Cognitive Empire: The Coming of Age of Epistemologies of the South.* Durham, NC: Duke University Press, 2018.

———. *Epistemologies of the South: Justice Against Epistemicide.* London: Routledge, 2014.

———. "Trump Won't Take Cyanide." *Critical Legal Thinking,* January 11, 2021. https://criticallegalthinking.com/2021/01/11/trump-wont-take-cyanide/.

Sassen, Saskia. *Expulsions: Brutality and Complexity in the Global Economy.* Cambridge, MA: Belknap, 2014.

Sattar, M.B. *Greed and Politics Is Burning Down the Lungs of the Earth: The Curious Case of Wildfires in Amazon Rainforest.* San Bernardino, CA: n.p., 2020.

Scheidel, Arnim, Daniela Del Bene, Juan Liu, Grettel Navas, Sara Mingorría, Federico Demaria, Sofía Avila, et al. "Environmental Conflicts and Defenders: A Global Overview." *Global Environmental Change* 63 (2020): 102–4.

Scheppele, Kim L. "Autocratic Legalism." *University of Chicago Law Review* 85, no. 2 (2018): 545–83.

Schuller, Mark. *Humanity's Last Stand: Confronting Global Catastrophe.* New Brunswick, NJ: Rutgers University Press, 2021.

Scott, Kim. *Benang: From the Heart.* Perth: Freemantle Press, 1999.

Scott, Rebecca R. *Removing Mountains: Extracting Nature and Identity from the Appalachian Coalfields.* Minneapolis: University of Minnesota Press, 2010.

Seager, Joni. *Carson's "Silent Spring": A Reader's Guide.* London: Bloomsbury, 2014.

Serhan, Yasmeen. "When the Far Right Picks Fights with a Teen." *Atlantic*, August 14, 2021. https://www.theatlantic.com/international/archive/2021/08/greta-thunberg-far-right-climate/619748/.

Shaxson, Nicholas. *The Finance Curse: How Global Finance Is Making Us All Poorer.* New York: Grove Atlantic, 2019.

Shiva, Vandana. *Making Peace with the Earth.* London: Pluto Press, 2013.

Slobodian, Quinn. *Globalists: The End of Empire and the Birth of Neoliberalism.* Cambridge, MA: Harvard University Press, 2018.

Steger, Manfred, and Amentahru Wahlrab. *What Is Global Studies? Theory and Practice.* London: Routledge, 2017.

Stiglitz, Joseph E. *Globalization and Its Discontents Revisited: Anti-Globalization in the Era of Trump.* New York: W.W. Norton, 2017.

Sze, Julie. *Environmental Justice in a Moment of Danger.* Oakland: University of California Press, 2020.

Talaga, Tanya. *All Our Relations: Finding the Path Forward.* n.p.: House of Anansi Press, 2018.

Tallbear, Kim. "Caretaking Relations, Not American Dreaming." *Kalfou: A Journal of Comparative and Relational Ethnic Studies* 6, no. 1 (2019): 24–41.

Tansel, Cemal Burak, ed. *States of Discipline: Authoritarian Neoliberalism and the Contested Reproduction of Capitalist Order*. London: Rowman & Littlefield, 2017.

Taylor, Dorceta E. *The Rise of the American Conservation Movement: Power, Privilege, and Environmental Protection*. Durham, NC: Duke University Press, 2016.

———. *Toxic Communities: Environmental Racism, Industrial Pollution, and Residential Mobility*. New York: New York University Press, 2014.

Teachout, Zephyr. *Break 'Em Up: Recovering Our Freedom from Big Ag, Big Tech, and Big Business*. New York: All Points Books, 2020.

Telesca, Jennifer E. *Red Gold: The Managed Extinction of the Giant Bluefin Tuna*. Minneapolis: University of Minnesota Press, 2020.

Thunberg, Greta. *Our House Is on Fire: Scenes of a Family and Planet in Crisis*. New York: Penguin Books, 2020.

Todd, Zoe. "Refracting the State Through Human-Fish Relations: Fishing, Indigenous Legal Orders, and Colonialism in North/Western Canada." *DIES: Decolonization, Indigeneity, Education, and Society* 7, no. 1 (2018): 60–75.

Torres, Mauricio, and Sue Branford. *Amazon Besieged: By Dams, Soya, Agribusiness, and Land-Grabbing*. Rugby, UK: Practical Action Publishing, 2018.

Tosun, Jale, and Marc Debus. "Right-Wing Populist Parties and Environmental Politics: Insights from the Austrian Freedom Party's Support for the Glyphosate Ban." *Environmental Politics* 30, nos. 1–2 (2021): 224–44.

Tricontinental. *Popular Agrarian Reform and the Struggle for Land in Brazil*. Tricontinental: Institute for Social Research, dossier no. 27, June 2020. Accessed September 1, 2020. https://www.thetricontinental.org/dossier-27-land/.

Tsing, Anna Lowenhaupt. *The Mushroom at the End of the World: On the Possibility of Life in Capitalist Ruins*. Princeton, NJ: Princeton University, 2017.

Uekoetter, Frank. *The Green and the Brown: A History of Conservation in Nazi Germany*. Cambridge: Cambridge University Press, 2006.

Veracini, Lorenzo. *Settler Colonialism: A Theoretical Overview*. New York: Palgrave Macmillan, 2010.

Vince, Gaia. *Adventures in the Anthropocene: A Journey to the Heart of the Planet We Made*. Minneapolis: Milkweed Editions, 2014.

Viscidi, Lisa, and Nate Graham. "Brazil Was a Global Leader on Climate Change. Now It's a Threat." *Foreign Policy*, January 4, 2019. https://foreignpolicy.com/2019/01/04/brazil-was-a-global-leader -on-climate-change-now-its-a-threat/.

Vitale, Alex S. *The End of Policing*. London: Verso, 2018.

Voyles, Traci Brynne. *Wastelanding: Legacies of Uranium Mining in Navajo Country*. Minneapolis: University of Minnesota Press, 2015.

Waldron, Ingrid R.G. *There's Something in the Water: Environmental Racism in Indigenous and Black Communities*. Halifax, NS: Fernwood Publishing, 2018.

Wallace-Wells, David. *The Uninhabitable Earth: Life After Warming*. New York: Tim Duggan Books, 2019.

Waltner-Toews, David. *On Pandemics: Deadly Diseases from Bubonic Plague to Coronavirus*. Vancouver, BC: Greystone Books, 2020.

Weir, Jessica K. "Lives in Connection." In *Manifesto for Living in the Anthropocene*, edited by Katherine Gibson, Deborah Bird Rose, and Ruth Fincher, 17–21. Brooklyn, NY: Punctum Books, 2015.

Wildcat, Daniel R. *Red Alert! Saving the Planet with Indigenous Knowledge*. Golden, CO: Fulcrum Publishing, 2009.

Wolfe, Patrick. "Settler Colonialism and the Elimination of the Native." *Journal of Genocide Research* 8, no. 4 (2006): 387–409.

Wolff, Michael. *The Man Who Owns the News: Inside the Secret World of Rupert Murdoch*. New York: Broadway Books, 2010.

Yusoff, Kathryn. *A Billion Black Anthropocenes or None*. Minneapolis: University of Minnesota Press, 2018.

Zumbansen, Peer, ed. *The Oxford Handbook of Transnational Law*. Oxford: Oxford University Press, 2021.

FURTHER READING

Neoliberalism and Its Attack on Democracy

Brown, Wendy. *In the Ruins of Neoliberalism: The Rise of Antidemocratic Politics in the West.* New York: Columbia University Press,
2019.

Chomsky, Noam. *Profit over People: Neoliberalism and Global Order.*
New York: Seven Stories Press, 1999.

Giroux, Henry A. *The Terror of Neoliberalism: Authoritarianism and
the Eclipse of Democracy.* Boulder, CO: Paradigm Publishers, 2004.

Harvey, David. *A Brief History of Neoliberalism.* Oxford: Oxford University Press, 2007.

Hickel, Jason. *The Divide: Global Inequality from Conquest to Free Markets.* New York: W.W. Norton, 2018.

Levitsky, Steven, and Daniel Ziblatt. *How Democracies Die.* New York:
Crown, 2018.

MacLean, Nancy. *Democracy in Chains: The Radical Right's Stealth
Plan for America.* New York: Penguin Books, 2018.

Oreskes, Naomi, and Erick M. Conway. *Merchants of Doubt: How
a Handful of Scientists Obscured the Truth on Issues from Tobacco
Smoke to Global Warming.* London: Bloomsbury Press, 2010.

Saad-Filho, Alfredo, and Lecio Morais. *Brazil: Neoliberalism Versus Democracy.* London: Pluto Press, 2018.

Slobodian, Quinn. *Globalists: The End of Empire and the Birth of Neoliberalism.* Cambridge, MA: Harvard University Press, 2018.

Authoritarianism and the Global Rise of the Radical Right

Applebaum, Anne. *Twilight of Democracy: The Seductive Lure of Authoritarianism.* New York: Doubleday, 2020.

Bello, Walden. *Counterrevolution: The Global Rise of the Far Right.* Black Point, NS: Fernwood Publishing, 2020.

Berberoglu, Berch. *The Global Rise of Authoritarianism in the 21st Century: Crisis of Neoliberal Globalization and the Nationalist Response.* London: Routledge, 2020.

Hochschild, Arlie Russell. *Strangers in Their Own Land: Anger and Mourning on the American Right.* New York: New Press, 2018.

Horesh, Theo. *The Fascism This Time: And the Global Future of Democracy.* Boulder, CO: Cosmopolis Press, 2020.

Jackson, Jack. *Law Without Future: Anti-Constitutional Politics and the American Right.* Philadelphia: University of Pennsylvania Press, 2019.

Mayer, Jane. *Dark Money: The Hidden History of the Billionaires Behind the Rise of the Radical Right.* New York: Penguin, 2017.

Prashad, Vijay, ed. *Strongmen: Trump-Modi-Erdogan-Duterte-Putin.* New Delhi: LeftWord, 2018.

Robinson, William I. *The Global Police State.* London: Pluto Press, 2020.

Tansel, Cemal Burak, ed. *States of Discipline: Authoritarian Neoliberalism and the Contested Reproduction of Capitalist Order.* London: Rowman & Littlefield, 2017.

Climate Crisis—Theories and Impacts

Foster, John Bellamy. *The Return of Nature: Socialism and Ecology.* New York: Monthly Review Press, 2020.

Kimmerer, Robin Wall. *Braiding Sweetgrass: Indigenous Wisdom, Scientific Knowledge, and the Teachings of Plants.* Minneapolis: Milkweed Editions, 2013.

Klein, Naomi. *This Changes Everything: Capitalism vs. the Climate.* New York: Simon & Schuster Paperbacks, 2014.

Kolbert, Elizabeth. *The Sixth Extinction: An Unnatural History.* New York: Picador, 2015.

Latour, Bruno. *Down to Earth: Politics in the New Climatic Regime.* Cambridge, UK: Polity, 2017.

Nixon, Rob. *Slow Violence and the Environmentalism of the Poor.* Cambridge, MA: Harvard University Press, 2011.

Schuller, Mark. *Humanity's Last Stand: Confronting Global Catastrophe.* New Brunswick, NJ: Rutgers University Press, 2021.

Shiva, Vandana. *Making Peace with the Earth.* London: Pluto Press, 2013.

Sze, Julie. *Environmental Justice in a Moment of Danger.* Oakland: University of California Press, 2020.

Thunberg, Greta. *Our House Is on Fire: Scenes of a Family and Planet in Crisis.* New York: Penguin Books, 2020.

Tsing, Anna Lowenhaupt. *The Mushroom at the End of the World: On the Possibility of Life in Capitalist Ruins.* Princeton, NJ: Princeton University, 2017.

Wallace-Wells, David. *The Uninhabitable Earth: Life After Warming.* New York: Tim Duggan Books, 2019.

Wildcat, Daniel R. *Red Alert! Saving the Planet with Indigenous Knowledge.* Golden, CO: Fulcrum Publishing, 2009.

Environmental Racism and Global Inequalities

Dowie, Mark. *Conservation Refugees: The Hundred-Year Conflict Between Global Conservation and Native Peoples.* Cambridge, MA: MIT Press, 2011.

Estes, Nick. *Our History Is the Future: Standing Rock Versus the Dakota Access Pipeline, and the Long Tradition of Indigenous Resistance.* London: Verso, 2019.

Gilio-Whitaker, Dina. *As Long as Grass Grows: The Indigenous Fight for Environmental Justice, from Colonization to Standing Rock.* Boston: Beacon Press, 2019.

LaDuke, Winona. *The Militarization of Indian Country.* East Lansing: Michigan State University Press, 2013.

Larsen, Soren C., and Jay T. Johnson. *Being Together in Place: Indigenous Coexistence in a More Than Human World.* Minneapolis: University of Minnesota Press, 2017.

Liberti, Stefano. *Land Grabbing: Journeys in the New Colonialism.* Translated by Enda Flannelly. London: Verso, 2013.

Talaga, Tanya. *All Our Relations: Finding the Path Forward.* n.p.: House of Anansi Press, 2018.

Taylor, Dorceta E. *Toxic Communities: Environmental Racism, Industrial Pollution, and Residential Mobility.* New York: New York University Press, 2014.

Waldron, Ingrid R.G. *There's Something in the Water: Environmental Racism in Indigenous and Black Communities.* Halifax, NS: Fernwood Publishing, 2018.

Yusoff, Kathryn. *A Billion Black Anthropocenes or None.* Minneapolis: University of Minnesota Press, 2018.

INDEX

Note: Page followed by "f" and "n" refer to figures and endnotes, respectively.